An English-Irish lexicon
of scientific and technical
space-related terminology

An English-Irish lexicon of scientific and technical space-related terminology

SUSAN McKENNA-LAWLOR
& DAMIEN Ó MUIRÍ

FOUR COURTS PRESS

Typeset in 10.5 pt on 14 pt Dante by
Carrigboy Typesetting Services for
FOUR COURTS PRESS LTD
7 Malpas Street, Dublin 8, Ireland
www.fourcourtspress.ie
and in North America for
FOUR COURTS PRESS
c/o ISBS, 920 NE 58th Avenue, Suite 300, Portland, OR 97213.

A catalogue record for this title is available
from the British Library.

ISBN 978–1–84682–269–8

Space Technology (Ireland) Ltd.

Printed in England
by MPG Books, Bodmin, Cornwall.

CONTENTS

NOTES ON THE AUTHORS

Professor Susan McKenna-Lawlor, DSc

Susan McKenna-Lawlor (née McKenna) was born in Dublin and carried out graduate and post-graduate studies at University College Dublin, the Dublin Institute for Advanced Studies and the University of Michigan. She was conferred with a PhD in Space Physics by the National University of Ireland (1976) and elected a Member of the Senate of NUI (1997–2002). Presently, she is Emeritus Professor at NUI Maynooth, Guest Professor at the Chinese Academy of Sciences, and Founder and Managing Director of Space Technology Ireland Ltd, a company that designs, tests and constructs instrumentation for space.

She has participated in the implementation of space experiments launched by the European Space Agency (ESA), the National Aeronautics and Space Administration (NASA), the Russian Space Agency, the Japanese Space Agency, the Chinese Space Agency and the Indian Space Agency.

In the capacity of Principal Investigator, she carried full scientific, technical and administrative responsibility for Ireland's EPONA experiment on the European Space Agency's Giotto mission, which encountered Halley's Comet (1986); made the first historic fly-by of Earth (1990) and encountered Comet Grigg-Skjellerup (1992). She also provided the national instrument LION for the European Space Agency's Solar Heliospheric Observatory (SOHO) and acted/acts as Co-Investigator for experiments on the Cluster (four-spacecraft) constellation, Mars Express, VENUS Express, SMART I (lunar) and Rosetta (Comet Churyumov Gerasimenko) missions.

She participated in NASA's Skylab, Solar Maximum, WIND and Gravity Probe B (Relativity) missions and also with the Russian Space Agency in the Phobos Mission to Mars and its Moons (2 instruments) and in the Mars 96 mission (4 instruments). In addition she participated in the Chinese Double Star Mission (as Principal Investigator for the NUADU experiment) and in the Indian Chandrayan Mission to the Moon (as a Co-Investigator for the SIR-2 experiment).

Under contract to ESA/ESTEC, her group made upgrades to the analogue electronics of the SST experiment on NASA's WIND spacecraft and also carried out a re-work of the COSPIN (SOLCON) Data Processing Unit for NASA's Space Shuttle Mission ATLAS. In addition, inputs were made for the Max Planck Institute (Lindau, Germany) to the design and construction of a Time of Flight (TOF) Mass Spectrometer which was subsequently flown on the Japanese GEOTAIL Mission.

She is presently a Leading Co-Investigator for the PICAM experiment, which is in preparation for launch on the ESA/Japanese Mission BepiColombo to planet Mercury and is engaged with the Swedish Institute of Space Physics in carrying out feasibility studies for an instrument under development for the proposed NASA/ESA Laplace Mission (to Jupiter).

She received an Irish Person of the Year Award and Associate Membership of the Royal Astronomical Society (1986); was awarded the Russian Tsiokovsky Gold Medal for 'Outstanding Contributions to Cosmonautics' (1988) and made an Honorary Citizen of San Jose, California, for technological achievement (1991). She received a presentation from ESA for her contribution to the successful encounter of spacecraft Giotto with comet Grigg-Skjellerup (1992) and was conferred with the Irish laureate Woman of Europe Award (1994) and with the Book Award of the International Academy of Astronautics (1998) for *Whatever Shines Should be Observed* (repr. 2003). She was awarded the Presidential Prize for Research Excellence by the National University of Ireland, Maynooth (2001) and received formal recognition from ESA for her important contributions to the ROSETTA Mission (2004). In addition, she participated in a Group Achievement Award from NASA for contributions to the success of the CLUSTER Mission in exploring geospace. She was the recipient of the Slovak Academy of Sciences International Prize 'in recognition of her outstanding scientific work', and was conferred with an honorary DSc for published work by the University of Ulster (2005). More recently, she participated in the Laurel Team Award of the International Academy of Astronautics for her contributions to the Double Star Mission (2010).

She is the author/co-author of more than 150 refereed publications on scientific and technical subjects and is a co-author of *Essential Spaceflight Dynamics and Magnetospherics* (2003). Also, she is: a member of the Editorial Board of the Space Weather International Journal; Co-Chair of 'Small Bodies Missions and Technologies' of the International

Astronautical Federation and Chair of 'Spacecraft Launcher and Aircraft Environments', which advises ESA on strategies and activities related to space weather. Professional memberships include: the International Academy of Astronautics; the International Astronomical Union; and Fellowship of the Institute of Physics (London).

Damien Ó Muirí, PhD

Damien Ó Muirí was born in Dublin and studied at University College Dublin and the School of Celtic Studies of the Dublin Institute for Advanced Studies. He obtained a BA in Celtic Studies, an MA in Modern Irish by major thesis and a PhD in Modern Irish. He was awarded the Mansion House Fund Scholarship in Irish (first place in the NUI) and the John Henry Curran Scholarship. Subsequently, he was awarded a Studentship in the School of Celtic Studies.

He went on to lecture in the Department of Modern Irish at St Patrick's College, Maynooth (which subsequently became the National University of Ireland, Maynooth). During his career there, he served as Dean of the Faculty of Celtic Studies, as Acting Head of the Department of Early and Mediaeval Irish and twice as Acting Head of the Department of Modern Irish, and was a member of the Academic Council on several occasions. He also served on Coiste Léann na Gaeilge of the Royal Irish Academy. He has published a book and many articles. He became an Accredited Translator under Foras na Gaeilge in 2006.

He studied law at the Honourable Society of King's Inns, Dublin, and was called to the Bar of Ireland in 1995. He was awarded an LLM from University College Dublin in 2002. He was called to the State Bar of California and the Federal Bar for the Central District of California in 2003. He is also a Fellow of the Chartered Institute of Arbitrators and an Accredited Mediator. He practices law in Dublin.

He is very interested in oriental matters. He ran extramural courses in Japanese Studies and in Oriental Studies in St Patrick's College. He was Vice-President of the Ireland-Japan Society, Director of its language school and twice Chairman of the Friends of the Chester Beatty Library.

FOREWORD

The International Academy of Astronautics (IAA) has been engaged in terminology and multilingual activities since its creation in 1960. Consequently, it is a most welcome initiative that a new language is now being added to the existing nearly twenty-five languages handled by the Academy.

Space activity is presently accessible to an expanding number of countries in all continents and this involves a huge number of languages. As a consequence, the language barrier should not be underestimated in any sustainable development program. As an example, the Academy is now heavily involved in Africa, with two permanent regional offices in Tunis and Yaoundé. Many African countries already have space programs of their own, or have plans to develop their role in space activity. These countries are using several hundred languages and thousands of dialects.

The publication of an English-Irish space lexicon is most welcome for several reasons: first, it celebrates the maturity of Ireland as a space-faring nation within the concert of European countries and the related European Space Agency. It also offers, from a cultural point of view, a wonderful opportunity to draw to the attention of the international space community the diversity of European culture, since the Irish language is often hidden by the English language. Indeed, Irish language and culture are transcending boundaries of countries and oceans and this work will be appreciated worldwide.

Dr Jean-Michel Contant
Secretary General, IAA
June 2010

PREFACE

The present English-Irish lexicon of over 3,500 astronautical terms is extracted from a twenty-language lexicon of the same terms published in 2010 by the International Academy of Astronautics on the occasion of its fiftieth anniversary. By way of introduction to this stand-alone version, we provide below information on the International Academy of Astronautics and on the work carried out over many years within the Academy to develop a multilingual lexicon of space-related terminology. Also discussed are: the astronomical heritage of Ireland; the Irish language; the method of translation adopted in preparing the English-Irish part of the Academy's lexicon; and some of the practical problems encountered in translating scientific and technical terms into Irish.

The International Academy of Astronautics

The International Academy of Astronautics (IAA) was founded in Stockholm on 16 August 1960 to: bring together on a regular basis the world's foremost experts in the discipline of astronautics; recognize the accomplishments of their peers; explore and discuss cutting-edge issues in space research and technology and provide direction and guidance in the non-military uses of space and the ongoing exploration of the Solar System.

The Academy, which initially was led by Dr Theodore von Karman (a pioneer in the evolution of rocketry, and the IAA's first president), is based on the tradition of the great classical scientific academies of the seventeenth century in Rome, London and Paris, which, in the earliest days of modern science, fostered scientific enquiry and the exchange of ideas and new information. In the words of its second president, Dr Frank J. Malina, 'the classical academies served in a remarkable manner the phenomenal advance of man's new method of understanding nature and of applying this understanding for the benefit of mankind'.

The purposes of the present IAA (which was instituted in an era when, for the first time in history, escape became possible from the bonds imposed on humans by gravity) are, as stated in the Academy statutes:

- To foster the development of astronautics for peaceful purposes;
- To recognize individuals who have distinguished themselves in a branch of science or technology-related to astronautics;
- To provide a program through which the membership can contribute to international endeavors and cooperation in the advancement of aerospace science, in cooperation with national science or engineering academies.

Recognition of the global significance of astronautics is inherent in the Academy, which is composed of members from all over the world (at present approximately sixty-five countries). Developments such as the easing of East-West tensions, the progressive integration of European economies, and the Asian economic revolution have in recent years enhanced political aspirations toward international cooperation in space. Cost, complexity and other pragmatic considerations dictate now, as never before, the necessity for cooperation among nations, and it is presently probable that such twenty-first century initiatives as the establishment of a manned lunar base and the launching of a manned mission to Mars will ultimately be realized as a consequence of international endeavors.

Internal structure of the IAA

In order to provide opportunities to: promote the initiation of focused projects within the Academy; facilitate and monitor such activities; enable the exchange of ideas and information among members and recommend to the Board of Trustees appropriate measures relating to the establishment of Program Committees and Study Groups, six Academy Commissions have been established:

- Commission I: Space Physical Sciences;
- Commission II: Space Life Sciences;
- Commission III: Space Technology and System Development;
- Commission IV: Space Systems, Operations and Utilization;
- Commission V: Space Policy, Law and Economics;
- Commission VI: Space and Society, Culture and Education.

The breadth of topics covered by these Commissions mirrors the diversity and complexity that characterizes astronautics itself and, in what follows, the terminology by means of which scientists and engineers working in these various regimes communicate (excluding those in Commission V), will be considered.

Terminology

In the process of studying nature, scientists encounter new phenomena, explore new locations, develop new concepts and techniques and proceed to give names to them. Most of these designations are known only to professionals; for instance the (sometimes vigorously contested) names given to newly recognized magnetospheric boundaries contained in specialized scientific publications. Due to advances in the popularization of science, however, some terms gradually become part of everyday language (such as Black Holes; the Termination Shock).

The discipline of managing terminology consists, among other things, of the following activities:

- analyzing the concepts and concept structures used in a field or domain of activity;
- identifying the terms assigned to the concepts;
- establishing correspondences between terms used in various languages;
- assembling newly created terminology in databases;
- managing terminology databases.

Over the fifty years since its foundation, specialists in astronautics working in the disciplines covered by various Academy Commissions have, in concert with their worldwide counterparts outside the Academy, coined and used thousands of words whereby evolving technological procedures and the scientific insights that result from their application can be expressed. Such new terms quickly appear in refereed publications and text books and are used at international conferences where they form part of the common linguistic currency whereby state-of-the-art ideas, processes and discoveries in astronautics are communicated.

History of studying space-related terminology within the IAA

Early in its history, it was recognized within the IAA that the burgeoning global development of a specialized technological and scientific vocabulary in the field of astronautics was a process that needed to be specially studied. In particular, it was recognized that a potential for *misunderstanding* as well as of *understanding* was inherent in the rapid development of such terminology, as associated concepts and ideas were passed from one language to another.

The first milestone in the activity of the IAA in this regard was the compilation of a seven-language *Astronautical Multilingual Dictionary of the IAA*, which was published by the Academy in Prague in 1970. This dictionary was dedicated to the memory of Theodore von Kármán, and its scientific editor was Prof. Rudolf Pesek.

Work on compiling a space dictionary can be divided into three stages, namely: (a) preparation of a 'lexicon' (a table of corresponding terms in different languages); (b) compilation of a 'dictionary' containing both terms and their definitions; and (c) configuration of a dictionary for machine translation (Matsukata et al., 2002). For the distribution of such information, paper media, electronic media (such as CD-ROM) and online systems employing a computer network may be used. The means of distribution most suitable for each stage were summarized by Matsukata et al. (2002) in a table which is reproduced here (Table 1).

Table 1 Comparison of distribution media (after Matsukata et al., 2002).

	LEXICON	DICTIONARY	MACHINE TRANSLATION
Paper media	Good	Better	Impossible
CD-ROM	Best	Best	Best
Online	Good	Better	Good

IAA Space Dictionary Edition Zero

In 1986, a project to create within the Academy a homogeneous and authentic space terminology data-base was initiated by Prof. Jacques-Louis Lions (President of the French Space Agency). In this connection, he set up an *IAA Multilingual Terminology Committee*, and interested members were invited to attend its first meeting on 8 April 1987. An important objective of the new committee (the participating members of which were called *Language Coordinators*) was the compilation of an English-French Basic List (BL) of astronautical terms.

Specialists in thirteen countries worked for several years on this task, using common worksheets. On 10 January 1990, attendees at the first *IAA Terminology Workshop* (held in Paris) reviewed the content of the, then available, list. After several iterations, a final list of 2,596 English terms was adopted and, based on its content, the IAA Secretariat published in paperback what was styled the *IAA Space Dictionary Edition Zero* as one of the contributions of the Academy to the *International Space Year.* This incorporated six bilingual (with associated trilingual) lexicons and it was circulated during the first *World Space Congress* in Washington (1992).

In the course of this congress, the first IAA *Symposium* on Multilingual Astronautical Terminology was held, and selected papers from the presentations made were subsequently published in a special issue devoted to *terminology* in the Academy journal, *Acta Astronautica* (see the preface to that volume by Guest Editor Iván Almár, 1992). It is noted that this Symposium initiated a tradition of mounting half-day and one-day meetings on terminology during a sequence of International Astronautical Congresses subsequently held at various locations.

The Multilingual Space Dictionary (Version 1.0)

In October 1994, S. Trutz, on behalf of the Hungarian 'Last Word Foundation' (LWF), informed the IAA Secretariat that he was ready to publish the available set of space dictionaries in computer-readable form using personal, state-of-the-art software. On 30 September 1995, Prof. Gerhard Haerendel, in his capacity as Vice-President for Publications, and Dr J.M. Contant, General Secretary of the IAA, signed an agreement with LWF for publication of the Dictionary. This version, which was entitled the *Multilingual Space Dictionary*, was published on two disks in 1996. It included fourteen languages (91 bilingual dictionaries, encompassing 182 dictionaries altogether, which translated from one language into another). This issue (Version 1.0) is sometimes also referred to as the *Multilingual Space Terminology Data Bank.*

Multilingual Space Dictionary (Versions 1.1, 1.2 and 1.2A)

In 1996, the Language Coordinators were requested to correct several misunderstandings in the Dictionary that had arisen due, among other things, to the ambiguity of some English terms. This corrected text, which was entitled *Version 1.1*, was demonstrated on two floppy disks in April 1996. In the same year, the Committee decided to combine recently available dictionaries in Arabic and Hindi with the material already to hand. The new issue (*Version 1.2*) was published in October 1996 by the LWF. The CD-ROM of this version, with the number of its terms slightly increased to 2,628, encompassed sixteen languages (120 bilingual dictionaries) and, renamed Version 1.2A, it was distributed by the IAA and the LWF to all participants at the Second *World Space Congress* in Houston, Texas (2002).

Multilingual Space Dictionary (Version 2.1)
Meanwhile within the Academy, in the course of general internal restructuring, the former Multilingual Terminology Data Base Committee was transformed into a Study Group of Commission VI (namely Study Group 6.1, with Professor Keiken Ninomiya as Chairman). To this group was assigned the task of developing a version of the Multilingual Space Dictionary that could be published as part of the celebrations attending the fiftieth anniversary of the foundation of the Academy in 2010. With this goal in view, the group concentrated in the first instance on increasing the overall number of languages. This action resulted in the compilation of what is referred to as Version 2.1, which, in addition to the sixteen languages formerly available, includes Irish, Greek, Ukrainian and Indonesian. Version 2.1 is already accessible on the internet (www.iaaweb.org), where it is provided with access software that runs on Microsoft Windows. Terms can be viewed in a multiplicity of languages if appropriate fonts are installed.

Multilingual Space Dictionary (Version 3.0)
Meanwhile, in response to a request made to the Chairpersons of relevant IAA Commissions that they might suggest additional words for the next edition, a list of over 1,000 words was received by Study Group 6.1. These were reviewed by the Study Group during the 2010 Spring Meeting of the IAA in Paris, when it was noted that a significant number of the words suggested were already included in Version 2.1. Ultimately, slightly fewer than 1,000 new words were selected for inclusion in Version 3.0.

Machine translation
It is anticipated that the electronic form of Version 3.0 will later be utilized for machine translation. This requires three kinds of inter-related information (Matsukata et al., 2002) namely:

- grammar (knowledge about the language);
- dictionary (knowledge about each word);
- general knowledge (field of specialization).

Several methodologies for machine translation are currently available, and discussions concerning the best solution for the present application are ongoing within the Academy, under the leadership of experts from the Japanese Space Agency.

An English-Irish lexicon of space terms

The decision to prepare an English-Irish lexicon of space terms as a contribution (originally) to Version 2.1 was made by one of the present authors (SMcK-L) against the background of ongoing Irish participation in the scientific and technical activities of the International Academy of Astronautics, combined with personal involvement in scientific missions launched by six major space agencies. In addition, Ireland has a heritage of astronomical observations that extends over more than 5,000 years and it thus seemed appropriate and timely to assemble a vocabulary which allows scientists and engineers in the present day to express contemporary issues in astronautics in the national language. In addition, it can be anticipated that translators, historians, educators, librarians and others working in a wide variety of fields, including the mass media, will find this collection of astronautical terms of interest and value.

With the intention of making the English-Irish lexicon more easily accessible to a wide range of users, the most recent edition was, with the permission of the IAA, separated into the present, stand-alone publication, while continuing in parallel to form an integral part of the multilingual dictionary of the IAA. In consequence of this heritage, the lexicon follows American-style spelling and adopts the numbering and word order used in Version 3.0.

The overall list of terms contains a number of English homographs. These are distinguished by a superscript number and guidance as to the meaning concerned is, in each case, given in round brackets. Also, in harmony with the style of the multilingual dictionary, in instances where there is more than one Irish term in use, a second/third Irish translation for a single English term is sometimes given.

Monitoring of solar cycles at Brú na Bóinne

Brú na Bóinne (Newgrange) is the largest member of a group of Neolithic passage graves located in the Boyne Valley, Co. Meath, about 50km from Dublin. This monument consists of a passage and inner chamber with three ornamented alcoves. A cairn of stones about 90m in diameter covers the passage and a kerb of massive slabs, some richly decorated, with their long edges touching, surrounds the base of the cairn and acts as a retaining element. A feature known as the 'roof box', which rests on the front part of the passage roof, is located some distance back from the entrance. It is 90cm wide, 1m high and 1.2m in depth and is situated above a gap 18–22cm wide between the first two roof slabs of the passage.

Radiocarbon dating provided by two charcoal samples from between the roof-slabs, when converted to a calendar date using the dendrochronological record provided by on-site samples of Irish oak, yields a probable construction date for the monument of 3,150±100BC. A study by Ray (1989) indicates that, at this period, the obliquity of the ecliptic would have been 24°2'±1', and the midwinter Sun would have risen at an azimuth of 133°54'±4' (to calculate the effect of refraction, a pressure of 1,010 mbar, a temperature of 0°C and a local horizon height of 55°±2' were adopted. Variations in the pertaining temperature and pressure, however, would make very little difference to the results obtained).

Under the circumstances of first construction, the Sun would have risen just within the limits of the roof-box as seen from the main chamber (the first-light axis of the passage corresponded, within 5' or so, to midwinter sunrise 5,150 years ago). If, at that time, Stone L20 of the monument, which is presently inclined, were in an upright position, the accuracy achieved would have been better than 25'. Overall, it can be deduced that, when Brú na Bóinne was built, the first beam of sunlight on the floor was less than 10cm wide and approximately 2m long, and that it bisected the chamber (Figure 1).

As the Sun's altitude and azimuth increased, the beam would gradually have become broader. The height of the gap behind the roof-box varies a little, but is about 20cm on average (see above). As seen from the main chamber, 18m away, this gap subtends an angle of roughly 38', so that the risen Sun would have been framed in the vertical direction (a color photograph of the appearance of the Sun at a recent winter solstice is shown on the dust jacket of this book).

At the time of construction of the monument, the lower solar edge would have exited the roof-box, as seen from the main chamber, at an azimuth of 137°42'±5'. This is very close to the maximum azimuth of the passage and implies symmetry with respect to the Sun's passage at the winter solstice, thereby indicating that not only the height but also the width of the gap is of significance (Ray, 1989).

This evidence supports the theory that the orientation of Brú na Bóinne (Newgrange) was deliberate, which makes it the oldest megalithic

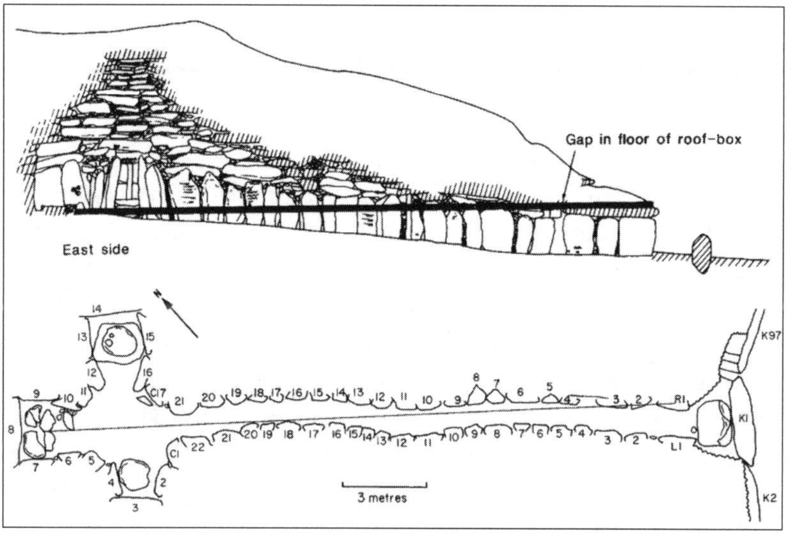

1 Sectional elevation of the east side of the Newgrange structure (above), showing the calculated path of the first rays of the Sun that entered the Roof Box at the time of the midwinter sunrise of 3150BC. Below: plan of the passage and main chamber at Newgrange, showing the minimum azimuth (or first-light) axis of the monument (after Ray 1989).

structure recognized to have an astronomical function. The monument pre-dates the astronomically orientated structures (Phase III) of Stonehenge by about 1,000 years and the Great Pyramid of Giza by about 600 years.

Record of lunar cycles at Cnobha (Knowth)

Evidence for the existence of an interest in lunar as well as in solar cycles in the Boyne Valley in Neolithic times was pointed out by Brennan (1994), who drew attention to patterns on a kerb-stone that forms part of another megalith in the southwest section of the site at *Cnobha* (Knowth). On this stone (Figure 2), there is a sequence of twenty nine symbols, one of which, immediately to the left of and above the large spiral, was interpreted by Brennan to represent the first appearance of the crescent of the waxing moon in the western sky. A lunar interpretation of the accompanying carvings appears to be plausible, since (a) there is a cyclic arrangement; (b) there is a transition from a crescent to a disk shape; and (c) the number of individual symbols depicted is twenty nine. Positions twenty-eight and twenty-nine are represented as hidden behind the spiral symbol. If the carvings, which date from about 3000BC, do indeed represent a lunar cycle, it is the earliest to be identified, without serious ambiguity, in either Ireland or Britain (McKenna-Lawlor, 1982).

Astronomy in pre-Christian/Early Christian Ireland

If we seek to determine information about astronomy in pre-Christian Ireland (Christianity was introduced in the fifth century), the earliest records available refer specifically to the Celts. It is a matter of scholarly debate as to when Celtic culture came to Ireland, but a very general date in the late Bronze Age of the order of 600BC can be adopted for general purposes. The testimony of classical Greek and Roman sources as well as Irish vernacular sources provide relevant information regarding the characteristics of Celtic society. In classical documents we find evidence of a high-ranking class of Celtic learned men who are referred to as Druids. Strabo mentions in particular druidical knowledge of moral philosophy and natural science, while Caesar, in his account of the conquest of Gaul, says of the druids:

> They hold long discussions about the heavenly bodies and their movements, the size of the universe and of the Earth, the physical constitution of the world and the power and properties of the Gods and they instruct the young men in all these subjects.

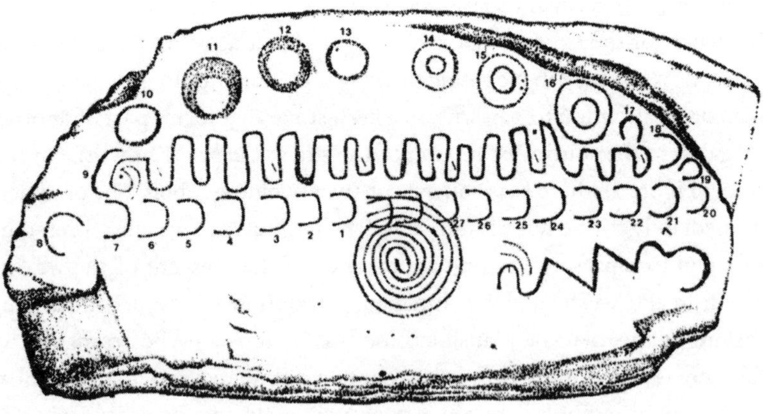

2 The 'calendar' stone at Knowth (from McKenna-Lawlor (1982)).

In addition, Pliny reported that the Druids measured their months and years by the Moon and also their ages of thirty years.

In Irish vernacular texts, although Druids are referred to, the ethnology of the word appears to be different, and there is a special emphasis in associating these personages with divination and with magic. The documents in addition identify certain orders of learned men that include, in particular, the *file*. The word *file* is made up of two elements, namely *fil*, which is cognate with the Welsh *gwel* (see) and the agental ending *-e*. It appears that, in its original meaning, *file* would have been similar in connotation to the English word 'seer'. Later, it came to be applied particularly to poets, although it had initially referred to a class whose members had wider functions. Of special interest is an incantation ascribed, from long before the Christian era, to the *file* Amerghin Gluingheal, which mentions specifically his knowledge of lunar and solar cycles:

> Who spreads light on the gathering on the mountain?
> Who foretells the ages of the Moon?
> Who teaches the spot where the Sun rests?
> Who, if it be not I?

While there is a question as to whether the *file* existed from earliest times side-by-side with the Druids (whom they came to survive), or if rather they represented part of an evolution within the druidical orders, history indicates that, at the time of the coming of Latin-based culture to Ireland in the fifth century, the Irish already had a long tradition of native learning which was taught in the druidical schools by the *file* through oral transmission. Thus, when pagan society was supplanted by Christianity and ecclesiastical schools took the place of those that had gone before, these new schools were internally stratified according to the ancient, so-called 'seven grades' or 'seven orders' of wisdom, whereby Professors of the fifth grade had, among other subjects, to be proficient in mathematics and astronomy.

The Irish language

The Irish language (which is sometimes referred to as Irish Gaelic in contrast to its sister language, Scottish Gaelic) is the first official language of Ireland (as the Republic of Ireland is now styled). It is spoken as a daily language in Gaeltachtaí (Irish-speaking areas), which are found in seven counties of the State and also by families and individuals outside of these locations. Irish is taught as a subject in every school and there are also Gaelscoileanna (schools where Irish is the only medium of instruction). There is an Irish language television station (TG4) and a radio station (Raidió na Gaeltachta), as well as weekly newspapers, magazines and books.

Irish is a very ancient language and is older than English (which supplanted British, a Celtic language that was the ancestor of Welsh). It is the third European language to be written down and, within Europe, only Latin and Greek are older written languages. It is an Indo-European language (one of the Celtic family of languages), but it is only within the last three hundred years that the Celtic origin of the peoples of Ireland and Britain was recognized and a connection made between the Goidelic branch (Irish, Scottish Gaelic and Manx) and the Brythonic branch (Welsh, Breton and Cornish). The former are classified as Q Celtic (the 'q' having become a 'c', which was always pronounced k, in antiquity) and the latter as P Celtic. For example, if one takes the word for 'everybody' in both Irish and Welsh, Irish *cách* [ka:x] (*cá* as in *caw*, rhyming with *jaw* and ch as in *loch*, *lough*) it is the exact equivalent of the Welsh *pawb* [paub] (*aw* as in *pow*, *wow*).

A feature much easier for a linguist or a layperson to recognize is the VSO (Verb, Subject Object) word-order structure which is unique to Celtic among Indo-European languages (for example, Irish *Fuair an múinteoir leabhar* [fu:ə r′ ə m′u:n′t′o:r′ l′aur] Welsh *Cafodd yr athro llyfr* [kavɔð ər aθro ə ɬivr] (Literally, 'Got the teacher a book', meaning The teacher got a book).

In addition, there is a palatal/non-palatal contrast between Irish consonants similar to the English s where *ship* contrasts with *sip*. Palatal consonants are marked conventionally with the superscript (′) in Irish phonetic transcription. Another feature is a system of initial mutations which is somewhat different in the two languages. In Irish, in the nominative-accusative singular case, masculine nouns being unlenited contrast with feminine nouns which are lenited, for example, *an fear mór* [ə f′ar mo:r] (on far more) 'the big man' and *an ghirseach mhór* [ə jirs′ax wo:r] (on yirshoch wore or vore) 'the big girl' and over 1,000 years ago both (in the nominative singular) would have contrasted with nasalized neuter *a n-ainm mór* [ə nan′əm′ mo:r] 'the big name'.

Modern Irish nouns, like those of German, have declensions, gender and case. On account of the fragmentation of the Irish speaking area, which extended at its height over all of Ireland and Scotland, there are now three distinct dialects of Modern Irish, namely Ulster (northern), Connacht (western) and Munster (southern) Irish. When all Ireland became fully subject to English rule in 1603, professional poets (bards) gradually disappeared from society and this loss marked the end of a standard literary language. Gaelic Ireland had no access at home to the printing press, and books published within Ireland tended to concern religious material from the Reformed Church (religious material inspired by the Counter-Reformation was meanwhile published in mainland Europe). Only with the Gaelic Revival in the late nineteenth century did modern Irish literature emerge.

The three dialects present a multiplicity of forms. For example, the same Irish word may have either three different pronunciations (rather like the English word *controversy*) or there may be three Irish words for a single English one. As an example, the Irish word for *salmon* is *bradán*, which is pronounced bradan in Ulster (and in Scottish Gaelic). It is brudawn in Connacht with the emphasis on the first syllable and bra dawn in Munster with the emphasis on the second syllable (which rhymes with Madonna on dropping the final *a*). On the other hand, there are three Irish words for *mackerel* from north to south *murlus(c)*, *ronnach* and *maicréal*. While one would expect the northern form, it is the western form which is found in Scottish Gaelic (*rionnach*).

An attempt has been made by the Government of Ireland to produce a standardized grammar and a reformed spelling system, and this is described in the publication *Gramadach na Gaeilge agus Litriú na Gaeilge An Caighdeán Oifigiúil* (Irish Grammar and Irish Spelling: the Official Standard). This standard, which was first published in 1958 and has been reprinted many times since then, was used as far as was practicable in preparing the present lexicon.

Translation method

The German version of the IAA Space Dictionary (Version 2.1) shows the gender of nouns and this practice has been followed in the present English-Irish translation using *m.* and *f.* (n. is not applicable). Where there were two or more English homonyms, the sequence adopted in the German lexicon in providing the differences in meaning was followed; for example, *hop* [1] and *hop* [2], where *hop* [2] means a short flight. For overall conformity, the United States spelling was also adopted.

There are two modern general dictionaries available in Irish, namely the *English-Irish Dictionary* edited by Tomás de Bhaldraithe (1959) and

Foclóir Gaeilge-Béarla (Irish-English Dictionary) edited by Niall Ó Dónaill, with consultant editor Tómas de Bhaldraithe (1977). Since both pre-date the advent of the internet, they have now become somewhat dated. The resulting void has been filled to some extent by the Focal.ie website. The www.focal.ie Project (2004–2007) was initiated by FIONTAR, DCU [the Fiontar interdisciplinary school of Dublin City University] in partnership with the University of Wales, Lampeter and Foras na Gaeilge (Institute of the Irish Language). The latter body, which is responsible for the promotion of the Irish language throughout the island of Ireland, was founded under the Good Friday Agreement as part of a North / South Implementation body set up to promote both the Irish language and the Ulster Scots language.

Focal.ie

There are over 150,000 English language terms at present in the Focal.ie terminology database (which is being continuously updated), drawn from a large number of domain-specific dictionaries and lists. Of these, the *Foclóir Réalteolaíochta / Dictionary of Astronomy* and the *Foclóir Eolaíochta / Dictionary of Science* in their original forms were particularly useful in the compilation of the present lexicon. Occasionally, a technical word, such as *bissextile* or *utricle*, was not found on the website but was contained in one of the above-mentioned general dictionaries. A term that surprisingly returned a negative result and only two related terms on the website is *propulsion*, for which *tiomáint* is given both there and in the general dictionaries.

At this point, it may be helpful to describe the method used in translating the terms. Typically, on typing an English (or an Irish) term in the search box of the fully bilingual Focal.ie website and clicking on search, the exact match(es) sought in English (or Irish) were returned and a new box containing a list of similar terms in English and / or Irish opened. If there was an exact match (or matches) for the English term, the website might return several Irish equivalents, listed according to the field within which they apply, accompanied by a list of related terms, usually word groups of two or more words beneath. For example, on typing the English term *range* and clicking on search, apart from opening a second box with seven similar terms in Irish and English, two exact English matches were returned, with one verb and one noun, the latter showing a list of thirteen Irish equivalents consisting of four different nouns and 112 related terms (the English term *range* appears in word groups, where it can occur in either initial, medial or final position in the group). Of course, on the other hand, the system may return Exact Matches (0) no terms found and Related Terms (0) no terms found, or one or more related terms.

If there is no term found, various strategies can be adopted. For example, if one types, in turn, the group terms *abdominal breathing*, *abdominal distension*, *aberration of light* the search returns a negative result for each group. However, if one types the single term *breathing*, the search lists many similar terms and returns one exact match, giving six Irish equivalents, comprising four different nouns, including *análú* as well as twenty related terms. Similarly, if one types the single term *abdominal*, the search returns two exact matches, one being an adjective giving the Irish equivalent as *boilg* viz. the genitive singular of the noun *bolg* 'abdomen', which is used as an adjective, or qualifier. Hence, one can combine the two terms to obtain an appropriate translation. On the other hand, if one types the English term *distension*, the search returns a negative result. Therefore, one needs to search for the headword *distension* in the *English-Irish Dictionary*, which yields *séideadh* and *borradh*. One then cross-checks the headwords for both *borradh* and *séideadh* in the *Foclóir Gaeilge-Béarla*. The headword *séideadh* is in five numbered parts where the five numbers signify a differentiation of meaning. At 3, *séideadh* is translated as 'inflation' and examples of usage are given, among others ~ *goile* distension of stomach (the symbol ~ stands for the whole headword when it is repeated in the explanatory paragraph). Again, typing *aberration of light* returns a negative result, whereas typing *aberration* on its own returns *aberration of starlight* among thirteen related terms, with its Irish equivalent *iomrall réaltsolais*. One need only then delete the first element in the compound to obtain an appropriate translation.

Again, the verb (to) *ablate* returns a negative result, but similar terms include *ablation*, which, if chosen, returns one exact match and two similar Irish terms *eisídiú* and *eisídiúchán*. Recognizing the former as a compound of *eis-* and *-ídiú*, one can turn to the *Irish-English Dictionary* and find several useful headwords to which *eis-* can be prefixed. As a final example, if one types *abort*, it returns one exact match and three Irish terms, but the one related term *abort a flight* contains the obviously appropriate term *tobscoir eitilt* and, returning to the three terms, one finds the verbal noun *tobscor*.

Some problems encountered

Even Focal.ie has its limitations, however, and the terminology committee and their staff recognizes these. Some terms have already become obsolescent and these are marked when they appear. Among the main kinds of questions (although no examples are given on the website) raised by both committee and users and which also were encountered in this work are:

How to differentiate between different concepts to which the same term has been designated

For example, the terms *aeronautics* and *air navigation* are both translated as *aerloingseoireacht*, probably because *aeronautics* is based on the Greek words meaning air navigation, but they have acquired different meanings in technical English. For the former, a new coining and a new combination *eitleolaíocht* f.; *eolaíocht* f. *eitilte* are given here.

Gravity has become associated with the Earth's gravity in Irish, but with the advent of space travel and space exploration, it has acquired a wider meaning. Therefore, *imtharraingt* is more appropriate than *domhantarraingt*, which is given as an alternative. The English words *collision* and *impact* can be translated as *imbhualadh*. They are better differentiated, however and so the usage in *Foclóir Eolaíochta/Dictionary of Science*, where *imbhualadh* is used for *collision* and *tuinseamh* is used for *impact*, is followed here. *Sensor* is translated by *brathadóir* in the field of astronomy, as is *detector*. However, a sensor is part of the detector and picks up the information which the detector brings to the attention of the human operator. Therefore, *braiteoir*, the term given for *sensor* in all other fields on the website, is used here. Another term, *ga*, is used for both *ray* and *radius* (which means ray in Latin), which could be confusing in space terminology, where both light and mathematics are important fields. For instance, *ga Schwarzschild* can mean either *Schwarzschild ray* or *Schwarzschild radius*, the latter being correct in our application. Therefore, a transliteration of *radius* (*raidias*) is used as an alternative here. Finally, as the word *domhain* 'deep' is identical in form to the genitive singular of *domhan* 'Earth', the word *cian* 'far, distant' was adopted to avoid confusion.

How to differentiate or choose between different terms designated to the same concept

There are frequently two or more words in Irish for the same concept. For example, Irish has two words for the English word *cloud*, namely *néal* and *scamall*, both in common use, with *scamall* occurring slightly more often on the website. There are three words for the English word *accident* – *timpiste*, *taisme* and *tionóisc*, all of which are in use.

Overenthusiastic differentiation can lead to more than one term. For example, the Irish word for *environment* was *timpeallacht*, but perhaps inspired by the German *umwelt*, *comhshaol*, a word found in the older *Foclóir Gaedhilge agus Béarla* (an Irish-English Dictionary) by Revd Patrick Dineen, meaning contemporary life, has been given a new connotation, with the meaning 'environment'. The compilers of different terminological dictionaries often appear to have worked independently, due perhaps to the long period over which these various works were produced. A recent example is the English term *feedback*, which is now translated by three Irish terms on the website (the bracketted words provided immediately below after these individual terms are the applicable domains accruing to particular fields), viz. *aischothú* (natural or cumulative process) for the fields of biology and the environment; *aisfhotha* (electronics) for the fields of science, computers/computer science, broadcasting/telecommunications/ telephone and electricity/electronics; and *aiseolas* (information) for the fields of sociology, administration and business. Two of these terms are found in the *Foclóir Gaeilge-Béarla: aiscothú* for the field of biology; and *aiseolas* as a general term, but only the single term *aischothú* is given in the *English-Irish Dictionary* for the fields of electrical engineering, wireless telephony and telegraphy. None of these words appear in Dinneen's Irish-English Dictionary. Other examples are the Irish words *feiste* and *gléas* for the English term *device*; *imeall* and *ciumhais* for *edge*; *seomra* and *cuasán* for *chamber* and *samhail*, *mionsamhail* and *múnla* for *model*.

One frequently encounters a verbal noun ending in *-(i)ú* or *-(e)adh*, for example *éilliú* 'contamination' and *craoladh* 'broadcast', and a noun based on the verbal noun with the nominal suffix *-chán*; for instance, *éilliúchán*; *craolachán*.

How to identify and deal with multiple versions of the same concept to which different terms have been designated (in both English and Irish)

The English terms *depression* and *low* and their Irish equivalents *ísleán* and *lagrach* provide an example of different terms designated to the same concept in both English and Irish.

It is noted that *Gyro* and *gyroscope* are invariably translated as *gíreascóp* here, there being no short form given on the website.

Compounds

In English, compound words are formed frequently from two elements (for example, *starlight*), whereas in Modern Irish the norm is to keep the two separate with the second element following the noun (*solas réaltaí* 'light of stars'). Prefixes (and indeed infixes), as in Modern German, were common in Old Irish (eighth/ninth century AD), but occur much less in spoken Modern Irish, although they have enjoyed a revival in literary Modern Irish. *Réaltsolas* 'starlight' would be a modern literary coining. Just as there is a difference between the English *a black bird* (any bird black in color) and *a blackbird* (specifically a member of the genus *turdus*), so too there is a difference between *lá gréine* 'a sunny day' and the technical term *grianlá* 'a solar day'. A solar panel, however, can be translated *painéal gréine* which corresponds to everyday speech, or *grianphainéal*, which is a literary formation. An attempt has been made here to standardize the usage by giving both forms as far as was practicable.

Prefixes

In the Space Lexicon, one frequently finds words with one or more prefixes. Sometimes the word in that form was not found in Focal.ie (for instance, *aerodyne* and *aerothermodynamics*, but it was then merely a matter of adding an (extra) existing prefix *aerai-* to an already existing word, such as *dín* (*dyne*) or *teirmidinimic* (*thermodynamics*), giving *aeraidín* and *aeraiteirmidinimic* respectively. Sometimes, two existing prefixes *aerai-* and *teirmi-* were needed (for example, *aeraiteirmileaisteachas* for the term *aerothermoelasticity*). In other examples, the word to which it was prefixed was not found (such as *(o)dontalgia* as in *aerodontalgia*). Here, the prefix *aerai-* was applied and the second element translated *déideadh* 'toothache', giving *aeraidéideadh*.

Older prefixes invariably cause mutations of the word to which they are prefixed (for example, mutation of *cobhsaí* 'stable' by lenition as in *sárchobhsaí* (< *sár* 'exceeding or ultra' + lenition *h*+ *cobhsaí*)) or rarely nasalization *éagobhsaí* (< *éa* negative 'in, un' + nasalizing *g*+*cobhsaí*) 'unstable'. Very recent modern prefixes which lenite the word to which they are prefixed are *geo-* in *geochobhsaí* 'geostationary' *meitea-* in *meiteachobhsaí* 'metastable' and *ultra* in *ultrachobhsaí* 'ultra-stable'. Some modern prefixes are listed on the website, but no examples of their usage with actual terms are given. Finally, contemporary use of the hyphen varies to some extent, as in English, with a trend away from hyphenation during the last century. *An Caighdeán Ofigiúil* (the Official Standard) gives guidance on the use of hyphens and, in addition, permits the use of a hyphen to avoid ambiguity.

Suffixes

The same situation can arise with suffixes as with prefixes, but to a lesser extent. For instance, an Irish equivalent for the term *mapper* was not found anywhere, and the agental ending *-ire* was added to *mapa* 'map' to form *mapaire*. Another word not found was *hypokinesy*, where *hipi* was prefixed to *cin* (as in *cinéitic* 'kinetics') and the word ending *-éis* (as found in the Irish word *cúirtéis* 'courtesy') added to form *hipicinéis*. Similarly, there was no word found for the important term *rocketry*, but by adding the abstract suffix *-acht* to *roicéad* 'rocket' a suitable term *roicéadacht* was formed.

Combinations of parts of two different words (portmanteau words)

Sometimes, two parts are taken from two different English words to form a new word, the so-called portmanteau words. For example, *azran* (<**az**imuth and **ran**ge); *ballute* (< **ball**oon and parach**ute**); *cermet* (<**cer**amic and **met**allic [material]); *companding* (<**com**pressing and ex**panding**) where the elements in bold type were combined to form new words. Firstly, each of these was transliterated giving *asran, balút, ceirmiot, compandáil* (*-áil* being a verbal suffix) and, secondly, the original two words were translated, giving *asamat agus raon; paraisiút balúin; ábhar ceirmeach miotalach; comhbhrú-fairsingiú*.

New or extended meanings

Some English words have acquired new or extended meanings, for instance *angel, dog-leg* (maneuver). These are translated firstly by their Irish equivalents: *aingeal; ionramháil cos con* and secondly by their meaning, *macalla paraisíteach (radair); ionramháil threoraithe chumachtaithe*.

The English word *coarsening* is a very technical term in space terminology, meaning an increase in the size of grains in a metal, and it is not found at all on the website. The general dictionary had no appropriate word that could be adapted, and therefore the meaning has been translated as *méadú gránmheide*. An easier term was *blockhouse*, which is primarily a military term but has been extended to NASA structures which protect observers from flying debris in case of catastrophic engine failure. The Irish word *dúnán* seemed appropriate and is used here.

Words not found

Needless to say, there were some terms which were not found, either on the website or in the general dictionaries (for example, *catergol* and *lithergol*, both of which are types of propellant). These have been transliterated as *catargal* and *liotargal* respectively, while *chuffing* is given as *tiufáil*, which was also assigned the word with the nearest sound (that is, *puthaíl*). A more difficult term was *sloshing*, a term used in fluid dynamics which refers specifically to the movement of propellants in spacecraft tanks and rockets, which ultimately was translated as *luascadh*.

Lenition of nouns followed by a qualifier

The nominative-accusative singular (dictionary form) of a noun can cause lenition of the initial consonant of a following qualifier, usually an adjective, or a noun in the genitive case. The initial consonants in question are *b, c, d, f, g, m, p, s* (except *sc, sm, sp, st* and the new *sf, sv*) and *t* marked in Roman type by a *h -bh, ch, dh, fh, gh, mh, ph, sh, th*. Initial *l, n, r* have never been marked for lenition in Irish spelling, but can be lenited in speech. Initial *h* in nouns occurs in later loan-words and is unaffected by the rule, as are modernisms using non-traditional *v* (as well as *q, x, y* and *z*).

Masculine nouns

Where the first noun is masculine in the nominative accusative singular (the dictionary form), no lenition occurs (for instance, *fear mór* 'big man'; *braiteoir gníomhach* 'active sensor'). Lenition does occur where masculine nouns in the nominative-accusative-dative plural end in a palatal consonant (such as *fir mhóra* 'big men', *caillteanais ghalaithe* 'boil-off losses').

Feminine nouns

Where the first noun is feminine and the qualifier is an adjective, lenition of a following adjective is the norm (*bean mhór* 'big woman'). The older non-lenition of dentals rule below is no longer followed: *bean dheas* '(a) nice lady'; *méid dhealraitheach* 'apparent magnitude' are usually written nowadays, rather than *d* without the *h*. The older non-lenited form *bean deas* is heard occasionally in spoken Irish.

Where the qualifier is a second noun (without an intervening definite article (feminine *na*)), however, the non-lenition of dentals (*d, s, t*) rule (that is, after a first noun ending in *d, l, n, s, t*) is followed here, as in Focal.ie, and also the non-lenition of *f*, a practice recommended by the Coiste Téarmáiochta and frequently found on the website, thus obviating the problem of whether to lenite or not to lenite (below). Examples of the applied rule are *seachmall dímheáchain* 'agravic illusion'; *scaird sainstiúrach* 'attitude jet'; *monatóireacht timpeallachta* 'environmental monitoring'. In all other cases, one would expect the initial consonant to be lenited, as with adjectives (above), but apart from a rather complicated list of exceptions giving situations where non-lenition occurs and contrasts with a situation where lenition occurs, contained in the *Caighdeán Oifigiúil* (Official Standard), there seems to be no obvious practice adopted yet on the website. An Coiste Téarmáiochta (the Terminological Committee) also has issued guidelines, but they, too, are complicated. It remains a grey area and poses a practical problem in a work like this, where there are so many new word groups, some of them quite novel. Such usage as could be ascertained through a diligent search of the website and the general dictionaries was followed.

The future

It is of course recognized that the present form of the English-Irish lexicon lacks an accompanying definition of the listed terms. Because of this, the meaning of some terms is ambiguous and, in these cases, synonyms will ultimately need to be identified and the order of their translation specified. Work to provide in English the formal definitions required to solve this inherent problem was started within the Academy in 2001 and is ongoing. It is anticipated by the present authors that future re-publication of the present lexicon with each term formally defined in both English and Irish will constitute the next step in allowing the lexicon to advance to the status of a dictionary.

In the past fifty years, for the first time in human history, it has become possible to build scientific instrumentation that can escape the bonds of gravity and fly to explore remote celestial objects. In-situ studies of distant planets and even of the edge of the Solar System are already ongoing and expeditions to colonize the Moon, Mars and beyond are under serious consideration. The possibility to seek for extraterrestrial life and to investigate profound questions such as how did the universe begin and how will it end have, in this ambience, not only become the province of specialists in the space agencies but are in addition openly discussed by the media and by the general public.

Given that a language is basically a system of communicating with others using sounds, symbols and words to express a meaning, idea or thought, the recent explosive development of scientific and technical knowledge over a plethora of disciplines requires from living languages the capability to rapidly evolve and provide to those who think and reason within their structures the tools to function in expanding intellectual areas. It is a pleasure to demonstrate through the present lexicon that the ancient, and very beautiful, Irish language possesses that capability.

Susan McKenna-Lawlor & Damien Ó Muirí
Maynooth, June 2010

REFERENCES

Almár, I. 2002. 'Astronautical terminology', *Acta Astronautica*, 50:2, pp 65–6.

Brennan, M. 1994. *The stones of time: calendars, sundials and stone chambers of ancient Ireland*. Rochester, VT.

De Bhaldraithe, T. 1959. *English-Irish Dictionary*. Dublin.

Dineen, P. 1927. *Foclóir Gaedhilge agus Béarla: an Irish-English dictionary*. Dublin.

Matsukata, J., T. Hashimoto, J. Kawaguchi, K. Ninomiya and R. Akiba. 2002. 'International edition of IAA multilingual dictionary coordinated through computer network', *Acta Astronautica*, 50:2, pp 95–101.

McKenna-Lawlor, S.M.P. 1984. 'Four thousand five hundred years of Irish astronomy' in M.A. Shea, D.F. Smart and S.M.P. McKenna-Lawlor (eds), *Proceedings of the Solar Terrestrial Interplanetary (STIP) Symposium on Solar Interplanetary Intervals at St Patrick's College, Maynooth, Ireland* (Chelsea, MI), pp 13–27.

Ó Dónaill, N. with consultant ed. T. de Bhaldraithe. 1977. *Foclóir Gaeilge-Béarla* (Irish-English dictionary). Dublin.

Ray, T.P. 1989. The winter solstice phenomenon at Newgrange, Ireland: accident or design, *Nature*, 337:6205, pp 343–5.

1958, repr. 2009. *Gramadach na Gaeilge agus litriú na Gaeilge: an caighdeán oifigiúil* (Irish grammar and Irish spelling: the official standard). Dublin.

1994. *Foclóir eolaíochta: Dictionary of science*. Enlarged ed., Dublin.

1996. *Foclóir réalteolaíochta: Dictionary of astronomy*. Dublin.

ABBREVIATIONS

f.	feminine	*mpl.*	masculine plural
fpl.	feminine plural	*m.*	masculine
pl.	plural	*s.*	singular

THE LEXICON

A

abdominal breathing análú *m.* boilg
abdominal distension séideadh *m.* boilg
aberration iomrall *m.*
aberration of light iomrall *m.* solais
ablate (to) eisídiú
ablating cone cón *m.* eisíditheach
ablating material ábhar *m.* eisíditheach
ablating nosecone soc-chón *m.* eisíditheach
ablation eisídiú *m.*
ablative eisídeach
ablative cooling fuarú *m.* eisídeach
abort tobscor *m.*
absolute gain dearbhneartúchán *m.*
absolute magnitude dearbhmhéid *f.*
absolute temperature dearbhtheocht *f.*
absolute zero dearbhniallas *m.*
absorptance ionsúiteas *m.*
absorption ionsú *m.*; ionsúchán *m.*
absorption band banda *m.* ionsúcháin
absorption factor fachtóir *m.* ionsúcháin
absorption line líne *f.* ionsúcháin
absorption spectrum speictream *m.* ionsúcháin
absorptive power cumhacht *f.* ionsúite
absorptivity ionsúiteacht *f.*
acceleration build-up méadú *m.* ar an luasghéarú
acceleration loading lódáil *f.* luasghéaraithe
accelerator luasaire *m.*
accelerometer méadar *m.* luasghéaraithe
acceptance glacadh *m.*
acceptance test tástáil *f.* ghlactha
acceptance test (to) glacadh *m.* a thástáil
access into space rochtain *f.* ar an spás
accident timpiste *f.*; taisme *f.*; tionóisc *f.*

acclimation clíomú *m.*
acclimatization cuibhiú *m.*
accommodation oiriúntas *m.*
accommodation coefficient comhéifeacht *f.* oiriúna
accretion fuilleamh *m.*; forthórmach *m.*
accretion disk diosca *m.* fuillimh; diosca *m.* forthórmaigh
accumulator taisc-cheallra *m.*
acoustic blanket blaincéad *m.* fuaim-mhaolaithe
acoustic delay line líne *f.* mhoillithe fuaime
acoustic generator gineadóir *m.* fuaime
acoustic radiation pressure brú *m.* radaíochta fuaime
acoustic vibration creathadh *m.* fuaime
acoustics fuaimíocht *f.*
acquisition fáil *f.* (comhartha)
acquisition and tracking radar radar *m.* fála agus rianaithe
actinometer achtanaiméadar *m.*
actinometry achtanaiméadracht *f.*
active gníomhach
active galactic nuclei núicléis *mpl.* réaltracha gníomhacha
active guidance treoraíocht *f.* ghníomhach
active homing frithing *f.* ghníomhach
active homing guidance treoraíocht *f.* frithinge gníomhaí
active rack isolation aonrú *m.* raca gníomhaigh
active redundancy iomarcaíocht *f.* ghníomhach
active satellite satailít *f.* ghníomhach
active sensor braiteoir *m.* gníomhach
actuator gníomhróir *m.*
adaptation phase pas *m.* oiriúnaithe
adapter cuibheoir *m.*
adapter skirt sciorta *m.* cuibheora
adiabatic wall temperature teocht *f.* balla aidiabataigh
adjustment coigeartú *m.*
administration riarachán *m.*
adsorption asú *m.*; asúchán *m.*
advanced propulsion ardtiomáint *f.*
advection trasiompar *m.*
aerial observation breathnóireacht *f.* aerga; breathnú *m.* aerga
aerial photography grianghrafadóireacht *f.* aerga

aerial reconnaissance taiscéalaíocht *f.* aerga
aerial remote sensing cianbhraiteacht *f.* aerga
aerodontalgia aeraidéideadh *m.*
aerodynamic braking coscánú *m.* aeraidinimiciúil
aerodynamic center lár *m.* aeraidinimiciúil
aerodynamic damping maolú *m.* aeraidinimiciúil
aerodynamic drag cúltarraingt *f.* aeraidinimiciúil
aerodynamic force fórsa *m.* aeraidinimiciúil
aerodynamic heating téamh *m.* aeraidinimiciúil
aerodynamic lift ardú *m.* aeraidinimiciúil
aerodynamic loads ualaí *mpl.* aeraidinimiciúla
aerodynamic rudder stiúir *f.* aeraidinimiciúil
aerodynamic stabilization cobhsú *m.* aeraidinimiciúil
aerodyne aeraidín *m.*
aeroelastic flutter eitleach *m.* aeraileaisteach
aeroelasticity aeraileaisteacht *f.*
aeroemphysema aereimfiséime *f.*
aerofoil section gearradh *m.* aereiteoige
aerolite aerailít *f.*
aerolith aerailit *f.*; aerailít *f.*; dreigít *f.* chlochach
aerology aereolaíocht *f.*
aeronautics eitleolaíocht *f.*; eolaíocht *f.* eitilte
aeronomy aeranomaíocht *f.*
aero-otitis media aeróitítis *f.* láir
aerosinusitis aeracuaisítis *f.*
aerosol aerasól *m.*
aerospace aeraspás *m.*
aerospace medicine leigheas *m.* aeraspáis; míochaine *f.* aeraspáis
aerospace remote sensing cianbhraiteacht *f.* aeraspáis
aerospace vehicle feithicil *f.* aeraspáis
aerospaceplane eitleán *m.* aeraspáis
aerosphere aeraisféar *m.*
aerostat aerastat *m.*
aerothermochemistry aeraiteirmiceimic *f.*
aerothermodynamic border teorainn *f.* aeraiteirmidinimiciúil
aerothermodynamics aeraiteirmidinimic *f.*
aerothermoelasticity aeraiteirmileaisteachas *m.*
aft dome cruinneachán *m.* cúil; cruinneachán *m.* deiridh

aft skirt sciorta *m.* cúil; sciorta *m.* deiridh
afterburning iardhó *m.*
aftercooling iarfhuarú *m.*
afterglow iarlaom *m.*; iarsholas *m.*
age of the moon aois *f.* na gealaí
agenda clár *m.*
aging aosú *m.*
agravic illusion seachmall *m.* dímheáchain; meascán *m.* mearaí de bharr dímheáchain
air aer *m.*
air hunger do-análú *m.*
air law dlí *m.* an aeir
air navigation aerloingseoireacht *f.*
air particles aercháithníní *mpl.*
air transportation aeriompar *m.*
air-breathing engine inneall *m.* aeranálaithe
aircraft aerárthach *m.*
airflight aereitilt *f.*
airflow aershreabhadh *m.*; aershreabh *f.*
airframe aerfhráma *m.*
airglow aerloinnir *f.*
airlock aerbhac *m.*
airplane experiment turgnamh *m.* eitleáin
airship aerlong *f.*
airshower aerchith *m.*
airsickness tinneas *m.* aeir; aerthinneas *m.*
airspace aerspás *m.*
Aitken nucleus núicléas *m.* Aitken
albedo ailbéideacht *f.*
albedometer méadar *m.* ailbéideachta
Alfven Mach number uimhir *f.* Alfvén Mach
Alfven speed luas *m.* Alfvén
Alfven waves tonnta *fpl.* Alfvén
aliasing ailiasáil *f.*
aligned currents srutha *mpl.* ailínithe
allocation of frequencies leithdháileadh *m.* minicíochtaí
all-or-nothing control rialú *m.* iomlán nó faic
all-sky search cuardach *m.* na spéire uile

allotment (**of frequencies**) cionroinnt *f.* (minicíochtaí)
almucantar almacantar *m.*
alpha decay alfa-mheath *m.*
alpha particle alfa-cháithnín *m.*
altimeter airdemhéadar *m.*
altitude airde *f.*
altitude acclimatization cuibhiú *m.* airde
altitude chamber seomra *m.* airde; cuasán *m.* airde
altitude measurement tomhas *m.* airde
altitude myopia gearr-radharc *m.* airde; míóipe *f.* airde
altitude sickness tinneas *m.* airde
ambient noise torann *m.* timpeallach
amino acid aimínaigéad *m.*
amnesia aimnéise *f.*
Amor asteroid astaróideach *m.* Amor
amplitude modulation [**AM**] modhnú *m.* aimplitiúide
amyloplast aimileaplast *m.*
anabolism anabalacht *f.*
analog-to-digital converter [**ADC**] tiontaire *m.* analóg go digiteach
anechoic chamber seomra *m.* neamh-mhacallach
anemia ainéime *f.*
angel aingeal *m.*; macalla *m.* paraisíteach (radair)
angle of attack uillinn *f.* ionsaithe
angle of climb uillinn *f.* dreaptha
angle of descent uillinn *f.* tuirlingthe
angle of elevation uillinn *f.* airde
angle of incidence uillinn *f.* teagmhasachta
angle of nutation uillinn *f.* nútaithe
angle of pitch uillinn *f.* chéime; céimuillinn *f.*
angle of precession uillinn *f.* luainíochta
angle of roll uillinn *f.* rolla
angle of yaw uillinn *f.* luascála
angle tracking rianú *m.* uilleann
angular diameter distance fad *m.* trastomhais uilligh
angular elevation airde *f.* uilleach
anisoelasticity ainiseaileaisteachas *m.*
announcement of opportunity fógairt *f.* deise
annual aberration iomrall *m.* bliantúil

annual parallax saobhdhiallas *m.* bliantúil
annular eclipse urú *m.* fáinneach; éiclips *m.* fáinneach
anomalistic month mí *f.* aimhrialta
anomalistic period tréimhse *f.* aimhrialta
anomalistic year bliain *f.* aimhrialta
anomaly aimhrialtacht *f.*
anoxia anocsacht *f.*
antapex frithrinn *f.*
antarctic meteorite dreigít *f.* antartach
antenna array eagar *m.* aerógach
antenna feed fotha *m.* aeróige
antenna gain neartú *m.* aeróige
antenna remote control cianrialú *m.* aeróige
antenna temperature teocht *f.* aeróige
anthropocentric antrapalárnach
anti-coincidence frith-chomhtheagmhas *m.*
anticyclone frith-chuaranfa *m.*
anti-g suit culaith *f.* fhrith-g; culaith *f.* frith-imtharraingthe
anti-gravity frith-imtharraingt *f.*; frithdhomhantarraingt *f.*
anti-gravity muscle matán *m.* frith-imtharraingthe
antimatter frithdhamhna *m.*
anti-orthostatism frith-ortastatachas *m.*
anti-sloshing frithluascadh *m.*
anti-sloshing baffle sciath *f.* frithluasctha
anti-vortex frithchuilithe *f.*
anxiety imní *f.*
apareon apaireon *m.*; fadphointe *m.* ó Mhars
apastron apastrón *m.*
aperture cró *m.*
apex rinn *f.*
aphelion aiféilean *m.*
apnea aipnia *f.*
apoapsis apaipsis *f.*; fadphointe *m.* fithise
apogee apaigí *m.*
apogee boost motor [**ABM**] mótar *m.* treisithe apaigí
apogee kick motor [**AKM**] cicmhótar *m.* apaigí
apogee kick stage cic-chéim *f.* apaigí
apogee motor mótar *m.* apaigí

Apollo asteroid astaróideach *m.* Apollo
apparent additional mass mais *f.* bhreise dhealraitheach
apparent eclipse urú *m.* dealraitheach; éiclips *m.* dealraitheach
apparent magnitude méid *f.* dhealraitheach
apparent sidereal time réaltam *m.* dealraitheach
appropriation leithghabháil *f.* bhuiséadach
approval test tástáil *f.* formheasa
apron naprún *m.*
apsis (*pl.* **apsides**) aipsis *f.* (*fpl.* aipsisí)
arcade formation foirmiú *m.* stuara
archaebacterium aircéabaictéir *m.*
arc-jet propulsion tiomáint *f.* stuascairde
area achar *m.*; limistéar *m.*
area expansion ratio cóimheas *m.* fairsingithe achair
area rule riail *f.* achair
areal velocity treoluas *m.* limistéarach
argument argóint *f.*
argument of latitude argóint *f.* leithid
argument of pericenter argóint *f.* peirealáir
argument of perigee argóint *f.* peirigí
arming armáil *f.*
array eagar *m.*
array (antenna) eagar *m.* (aerógach)
array of solar cells eagar *m.* griancheall
arrow wing sciathán *m.* saighde
artery heat pipe píobán *m.* artaire teasa
artificial aurora experiment turgnamh *m.* gealáin shaorga
artificial gravity imtharraingt *f.* shaorga
artificial horizon léaslíne *f.* shaorga
artificial intelligence intleacht *f.* shaorga [IS]
artificial respiration riospráid *f.* shaorga
artificial satellite satailít *f.* shaorga
ascending node nód *m.* éiritheach
ascent éirí *m.*
ascent path conair *f.* éirí
assembly building foirgneamh *m.* cóimeála
assignment (of frequencies) leithdháileadh *m.*; dáileadh *m.* (minicíochtaí)

assimilation comhshamhlú *m.*
assistance cúnamh *m.*; cabhair *f.*; cuidiú *m.*
assisted take-off éirí *m.* cuidithe
association[1] **(stellar)** réaltghaolmhaireacht *f.*; gaolmhaireacht *f.* réaltach
association[2] **(society)** cumann *m.*
asteroid astaróideach *m.*; mionphláinéad *m.*; pláinéadóideach *m.*
astrionics leictreonaic *f.* spásaireachta
astrobiology réalt-bhitheolaíocht *f.*
astrocompass réaltchompás *m.*
astrodynamics astraidinimic *f.*
astrolabe astraláib *f.*
astrometric position suíomh *m.* réaltmhéadrach
astrometry réaltmhéadracht *f.*
astronaut spásaire *m.*
astronautical spásaireachta
astronautical law dlí *m.* na spásaireachta
astronautical velocity treoluas *m.* spásaireachta
astronautics spásaireacht *f.*
astronavigation réaltloingseoireacht *f.*
astronomical almanac almanag *m.* réalteolaíoch
astronomical constant tairiseach *m.* réalteolaíoch
astronomical coordinate comhordanáid *f.* réalteolaíoch
astronomical coordinate system córas *m.* comhordanáideach réalteolaíoch
astronomical day lá *m.* réalteolaíoch
astronomical equator meánchiorcal *m.* réalteolaíoch
astronomical latitude leithead *m.* réalteolaíoch
astronomical longitude fad *m.* réalteolaíoch
astronomical meridian fadlíne *f.* réalteolaíoch
astronomical parallel líne *f.* chomhthreomhar réalteolaíoch
astronomical time am *m.* réalteolaíoch
astronomical twilight breacsholas *m.* réalteolaíoch
astronomical unit [AU] aonad *m.* réalteolaíoch [AR]
astronomical zenith forar *m.* réalteolaíoch
ataxia neamhluail *f.*
Aten asteroid astaróideach *m.* Aten
athermanous teasteimhneach
atmosphere atmaisféar *m.*

atmospheric absorption coefficient comhéifeacht *f.* ionsúcháin atmaisféaraigh
atmospheric braking coscánú *m.* atmaisféarach
atmospheric disturbance suaitheadh *m.* atmaisféarach
atmospheric electricity leictreachas *m.* atmaisféarach
atmospheric entry iontráil *f.* atmaisféarach; iontráil *f.* san atmaisféar
atmospheric mass mais *f.* atmaisféarach
atmospheric noise torann *m.* atmaisféarach
atmospheric optics optaic *f.* atmaisféarach
atmospheric pressure brú *m.* atmaisféarach
atmospheric radiation radaíocht *f.* atmaisféarach
atmospheric refraction athraonadh *m.* atmaisféarach
atmospheric scrubber sciúrthóir *m.* atmaisféarach
atmospheric sounding sondáil *f.* atmaisféarach
atmospheric tide taoide *f.* atmaisféarach
atmospheric transmissivity tarchurthacht *f.* atmaisféarach
atmospherics aerthormán *m.*
atomic clock clog *m.* adamhach
atomic gyro gíreascóp *m.* adamhach
atomic radiation radaíocht *f.* adamhach
atrophy seargadh *m.*; meath *m.*
attenuation maolú *m.*; maolúchán *m.*
attenuation constant tairiseach *m.* maolaithe
attitude sainstiúir *f.*
attitude and orbit control system [AOCS] córas *m.* rialaithe fithise agus sainstiúrach
attitude control rialú *m.* sainstiúrach
attitude control system [ACS] córas *m.* rialaithe sainstiúrach
attitude control unit [ACU] aonad *m.* rialaithe sainstiúrach
attitude gyro gíreascóp *m.* sainstiúrach
attitude jet scaird *f.* sainstiúrach
attitude sensor braiteoir *m.* sainstiúrach
attitude-stabilized satellite satailít *f.* sainstiúrach cobhsaithe
attribute testing tástáil *f.* tréithe
audiometry closmhéadracht *f.*
auditory sensation area achar *m.* mothúcháin éistithigh
aurora gealán *m.*
aurora australis na gealáin *mpl.* theas

aurora borealis na gealáin *mpl.* thuaidh; na saighneáin *mpl.*; an chaor *f.* aduaidh
auroral oval ubhchruth *m.* gealánach
auroral zone són *m.* gealánach
auto-igniting propellant tiomántán *m.* uathadhainteach
auto-ignition temperature teocht *f.* uathadhainte
auto-inflation uathshéideadh *m.*
automatic celestial navigation loingseoireacht *f.* réalteolaíoch uathoibríoch
automatic frequency control uathrialú *m.* minicíochta
automatic gain control [AGC] uathrialú *m.* neartúcháin
automatic pilot uathstiúir *f.*
automatic tracking rianú *m.* uathoibríoch
autopilot unit aonad *m.* uathstiúrach
auto-tracking uathrianú *m.*
auxiliary fluid ignition adhaint *f.* sreabháin chúnta
auxiliary landing gear gléasra *m.* tuirlingthe cúnta
auxiliary power unit [APU] aonad *m.* cumhachta cúnta
auxin polarity movement gluaiseacht *f.* pholach áicsin
availability infhaighteacht *f.*
avalanche photodiode [APD] fótaidhé-óid *f.* mhaidhme
aviation eitlíocht *f.*
avionics eitleonaic *f.*
axial-flow compressor comhbhrúiteor *m.* ais-sreafa
Az-El mount taca *m.* Az-El; taca *m.* asamait-airde
azimuth asamat *m.*
azimuth mount taca *m.* asamait
azran asran *m.*; asamat *m.* agus raon

B

babble gleoiseadh *m.*
backfire antenna aeróg *f.* chúltorta
background noise torann *m.* cúlra
background radiation radaíocht *f.* chúlra(ch)
back-scattering cúlscaipeadh *m.*
back-scattering coefficient comhéifeacht *f.* chúlscaipthe
backup cúltaca *m.*
backward contamination cúléilliú *m.*
bacteria baictéir *m.*
baffle sciath *f.*
bailout bottle buidéal *m.* ocsaigine (pearsanta)
bakeout bácáil *f.* amach; díghasáil *f.* tré théamh
balancing cothromú *m.*
ball lightning caorthintreach *f.*
ballast ballasta *m.*
ballistic balaistíoch
ballistic camera ceamara *m.* balaistíoch
ballistic flight eitilt *f.* bhalaistíoch
ballistic missile diúracán *m.* bailistíoch
ballistic re-entry athiontráil *f.* bhailistíoch
balloon balún *m.*
balloon launch lainseáil *f.* balúin
balloon release scaoileadh *m.* balúin
balloon release station stáisiún *m.* scaoilte balúin
ballute balút *m.*; paraisiút *m.* balúin
band banda *m.*
band-pass filter scagaire *m.* bandabhealaigh
bang-bang control rialú *m.* ann as
baroreceptor reflexes athfhillteacha *mpl.* baraghabdhdáin
barycenter baralár *m.*
base bun *m.*; bonn *m.*
base connector bun-nascaire *m.*

base drag bun-chúltarraingt *f.*
base pressure bunbhrú *m.*
baseband bunbhanda *m.*
baseline bunlíne *f.*; bonnlíne *f.*
basic maximum usable frequency bun-uasmhinicíocht *f.* inúsáidte
basket bascaed *m.*
batch processing baiscphróiseáil *f.*
baud bád *m.*
bay bá *f.*
beacon rabhchán *m.*
beam léas *m.*
beam footprint lorg *m.* léis
bearings and power transfer assembly [**BAPTA**] cóimeáil *f.* aistrithe cumhachta agus imthacaí
bed rest study staidéar *m.* taca droma
beginning of life [**BOL**] tús *m.* na beatha
behavior iompar *m.*
bell-shaped nozzle soc *m.* clogchruthach
Belousov-Zhabotinsky reaction imoibriú *m.* Belousov-Zhabotinsky
belt crios *m.*
beta decay béite-mheath *m.*
beta particle béite-cháithnín *m.*
bid tairiscint *f.*
bidder tairgeoir *m.*
Big Bang Ollphléasc *f.*
big-bang nucleosynthesis núicléisintéis *f.* ollphléisce
binary digit digit *f.* dhénártha
binary star réalta *f.* dhénártha
binder ceanglóir *m.*
bio-astronautics bithspásaireacht *f.*
bio-astronomy bithréalteolaíocht *f.*
bio-barrier bithbhacainn *f.*
bio-box bithbhosca *m.*
bio-burden reduction laghdú *m.* bithualaigh
bio-lab bithshaotharlann *f.*
biological isolation aonrú *m.* bitheolaíoch
bio-markers bithmharcóirí *mpl.*
bionics bitheoinic *f.*

bio-pack bithphaca *m.*
bio-pan bithphanna *m.*
bio-rack bithraca *m.*
bio-reactor bith-imoibreoir *m.*
bio-reduction strategies straitéisí *fpl.* bithlaghdaithe
bio-satellite bithshatailít *f.*
bio-shield bithsciath *f.*
bio-signature bithshíniú *m.*
biosphere bithsféar *m.*
bio-stack bithstaca *m.*
bio-telemetry biteiliméadracht *f.*
biotron biotrón *m.*
bipolar flow sreabhadh *m.* dépholach; sreabh *f.* dhépholach
bipropellant déthiomántán *m.*
bissextile year bliain *f.* bhisigh
bistatic radar radar *m.* déstatach
bit giotán *m.*
bit error rate [BER] ráta *m.* earráidí giotáin
black-body dúchorp *m.*
black-body radiation radaíocht *f.* dúchoirp
black box bosca *m.* dubh
black hole poll *m.* dubh; dúpholl *m.*
black-hole evaporation galú *m.* dúphoill
blackout lándorchú *m.*
blind flying eitilt *f.* dhall; dalleitilt *f.*
block diagram blocléaráid *f.*
block-house dúnán *m.*
blow-off séideadh *m.* amach
blunt body cabhail *f.* mhaol; corp *m.* maol
boat-tail deireadh *m.* báid
body cabhail *f.*; corp *m.*
body mass loss caillteanas *m.* corpmhaise
body-stabilized cabhailchobhsaithe
boil-off galú *m.*
boil-off losses caillteanais *mpl.* ghalaithe
bolide bóilíd *f.*
bolometer bólaiméadar *m.*
Bond albedo ailbéideacht *f.* Bond

bonding nascadh *m.*
bone density cnámhdhlús *m.*; dlús *m.* cnáimhe
bone loss caillteanas *m.* cnáimhe
bone matrix protein próitéin *f.* maitríse cnáimhe
bone metabolism meitibileacht *f.* chnáimhe
bone mineral density dlús *m.* mianra cnáimhe
bone resorption athshú *m.* cnáimhe
booster treiseoir *m.*
boot-strap system córas *m.* bútála
Bose-Einstein condensation comhdhlúthú *m.* Bose-Einstein
bottom bun *m.*; íochtar *m.*
bottomside (ionospheric) sounding sondáil *f.* (ianaisféarach) íochtair
Bouguer anomaly aimhrialtacht *f.* Bouguer
bounce boc *m.*; bocáil *f.*
bounce period tréimhse *f.* bhocála
boundary teorainn *f.*
boundary layer sraith *f.* teorann; ciseal *m.* teorann
Boussinesq approximation neastachán *m.* Boussinesq
bowser tancaer *m.*
bowshock turraing *f.* thosaigh
breadboard (model) losaid *f.* (samhail)
break-off scaradh *m.*
break-off phenomenon feiniméan *m.* scartha
breathing vibration creathadh *m.* anála
bremsstrahlung bremsstrahlung *m.*; radaíocht *f.* choscánaithe
brief (to) faisnéis *f.* a thabhairt
briefing seisiún *m.* faisnéise
brightness gile *f.*
broad line region réigiún *m.* leathanlíne
broadcasting craoladh *m.*; craolachán *m.*
broadcasting-satellite service seirbhís *f.* satailíte craolacháin
brown dwarf abhac *m.* donn
bubble boilgeog *f.*
bucket-brigade device feiste *f.* buicéad; gléas *m.* buicéad
buckling búcláil *f.*
buffer maolán *m.*
buffeting tuairteáil *f.*
bug fabht *m.*

building blocks of life bloic *mpl.* thógála na beatha
bulge bolg *m.*
bulkhead bulcaid *f.*
bungee cords cordaí *mpl.* buinsí
buoyancy convection comhiompar *m.* buacachta
buoyant convection comhiompar *m.* buacach
buoyant force fórsa *m.* buacach
burn rate ráta *m.* dó
burn-in (testing) dó *m.* isteach; triailrith *m.* (tástáil)
burn-out dó *m.* iomlán
burning dó *m.*
burning surface dromchla *m.* dó
burning surface to throat area ratio cóimheas *m.* an dromchla dó le hachar na scornaí
burning time (of an engine) am *m.* dó (innill)
bus bus *m.*
busbar busbarra *m.*
byte beart *m.*

C

cable cábla *m.*
cable-cutter gearrthóir *m.* cábla
caging blocáil *f.*
caisson disease galar *m.* casúin; arraingeacha *fpl.* díbhrú
calcium balance cothromaíocht *f.* chailciam
calcium distribution dáileadh *m.* cailciam
calcium excretion eisfhearadh *m.* cailciam
calcium loss caillteanas *m.* cailciam
calcium metabolism meitibileacht *f.* chailciam
calendar year bliain *f.* féilire
calibration calabrú *m.*; calabrúchán *m.*
call for tenders glao *m.* ar thairiscintí
call number gairmuimhir *f.*
canal-otolith interaction idirghníomhaíocht *f.* chanáil-otailite
capsule capsúl *m.*
capture gabháil *f.*
carbon star réalta *f.* charbóin
carbonaceous chondrite coindrít *f.* charbónúil
carbonaceous meteorite dreigít *f.* charbónúil
cardiac atrophy seargadh *m.* cairdiach; meath *m.* cairdiach
cardiovascular deconditioning díriochtú *m.* cardashoithíoch
cardiovascular system córas *m.* cardashoithíoch
cargo lasta *m.*
cargo bay bá *f.* lasta
carrier iompróir *m.*
case cás *m.*
Cassegrain system córas *m.* Cassegrain
Cassegrain telescope teileascóp *m.* Cassegrain
Cassini's law dlí *m.* Cassini
casting teilgean *m.*
cataclysmic variable star réalta *f.* athraitheach mhórchorraíola
catastrophic failure teip *f.* thubaisteach

catergol catargal *m.*
cavitation cuasú *m.*
ceiling síleáil *f.*
ceiling altitude uasairde *f.*; airde *f.* síleála
ceilometer síleáilmhéadar *m.*
celestial body rinn *m.* neimhe
celestial coordinates comhordanáidí *fpl.* neamhaí
celestial longitude fad *m.* neamhaí
celestial mechanics meicnic *f.* neamhaí
celestial meridian fadlíne *f.* neamhaí
celestial object réad *m.* neamhaí
celestial sphere sféar *m.* neamhaí
cell cill *f.*
censured test tástáil *f.* cháinte
Centaur (minor planet) ceinteár *m.* (mionphláinéad)
center of gravity meáchanlár *m.*; lár *m.* meáchain
center of inertia lár *m.* táimhe
center of mass maislár *m.*; lár *m.* maise
center of pressure brúlár *m.*; lár *m.* brú
center of thrust lár *m.* sá
centerbody nozzle soc *m.* lárchabhlach
central meridian fadlíne *f.* lárnach
central processing unit [CPU] láraonad *m.* próiseála [LAP]
centrifugal compressor comhbhrúiteoir *m.* lártheifeach
centrifuge lártheifneoir *m.*
centrifuge accommodation module [CAM] modúl oiriúntais lártheifneora
Cepheid variable star réalta *f.* athraitheach Cepheid
cermet ceirmiot *m.*; ábhar *m.* ceirmeach miotalach
certificate teastas *m.*; teistiméireacht *f.*; deimhniú *m.*
certification deimhniú *m.*; deimhniúchán *m.*
certification test tástáil *f.* deimhniúcháin
chain slabhra *m.*
chalcophile element eilimint *f.* chalcaifileach
chamber pressure brú *m.* seomra; brú *m.* cuasáin
Chandler period tréimhse *f.* Chandler
channel cainéal *m.*
characteristic exhaust velocity treoluas *m.* sceite saintréitheach

characteristic length fad *m.* saintréitheach
characteristic velocity treoluas *m.* saintréitheach
charge-coupled device [**CCD**] gléas *m.* luchtchúpláilte
charge-transfer device [**CTD**] feiste *f.* luchtaistriúcháin; gléas *m.* luchtaistriúcháin
charring ablator eisíditheoir *m.* gualaitheach
checkout fiorú *m.*; seiceáil *f.* amach
check out (**to**) lánseiceáil; seiceáil amach
checkout equipment trealamh *m.* seiceála amach
checkout system córas *m.* seiceála amach
check-up scrúdú *m.*; seiceáil *f.*
chemical evolution éabhlóid *f.* cheimiceach
chemical propulsion tiomáint *f.* cheimiceach
chemical radio-protector radachosantóir *m.* ceimiceach
chicken legs cosa *fpl.* sicín; tanú *m.* colpa
chill-down réamhfhuarú *m.*
chilling cuisniú *m.*
chip slis *f.*
chirality ciriúlacht *f.*
chirp modulation modhnú *m.* gíog; modhnú *m.* minicíochta líneach
choking Mach number uimhir *f.* thachtach Mach
Cholodny-Went theory teoiric *f.* Cholodny-Went
chondrite crondaít *f.*
chromosomal aberrations iomraill *mpl.* chrómasómacha
chromosphere crómaisféar *m.*
chromospheric plage scead *f.* chrómaisféarach
chuffing tiufáil *f.*; puthaíl *f.*
cigarette-burning dó *m.* toitín; dó *m.* tosaigh
circadian rhythm rithim *f.* laethúil
circular orbit fithis *f.* chiorclach
circular velocity treoluas *m.* ciorclach
circumnutation tiomghabháil *f.*
civil twilight breacsholas *m.* oifigiúil
clamshell fairing slíomú *m.* sliogáin brealláin
Clarke orbit fithis *f.* Clarke
class 0 protostar prótairéalta *f.* aicme 0
clean room seomra *m.* glan
cleanliness class aicme *f.* glaineachta

clean-up glanadh *m.* (suas)
climate aeráid *f.*
climatic test tástáil *f.* aeráide
climatology clíomeolaíocht *f.*
clinostat clíneastat *m.*
closed ecological life support systems [CELSS] córais *mpl.* tacaíochta beatha éiceolaíche dúnta; éiceachórais *mpl.* tacaíochta beatha dúnta
closed ecological system córas *m.* dúnta éiceolaíoch; éiceachóras *m.* dúnta
closed universe cruinne *f.* dhúnta
closing rate luas *m.* druidime; ráta *m.* dúnta
cloud néal *m.*; scamall *m.*
cloud height meter méadar *m.* airde néalta
cloud physics fisic *f.* néalta
cloud seeding síolú *m.* néalta
cluster braisle *f.*
cluster of galaxies braisle *f.* réaltraí
clutter cliotaráil *f.*; comharthaí *mpl.* paraisíteacha
co-located satellites satailítí *fpl.* comhshuite
coarsening méadú *m.* gránmhéide
coast[1] **(shore)** cósta *m.*
coast[2] **(moving without maintaining use of propelling power)** cóstóireacht *f.*; eitilt *f.* gan chumhacht
coastal zone color scanner [CZCS] scanóir *m.* datha són cósta
coasting flight eitilt *f.* chóstóireachta
coasting phase pas *m.* cóstóireachta
code cód *m.*
code conversion tiontú *m.* cóid
code converter tiontaire *m.* cóid
code-division multiple access [CDMA] ilrochtain *f.* chódroinnte
coding códú *m.*
coefficient comhéifeacht *f.*
cognitive decrements in space deicrimintí *fpl.* cognaíocha sa spás
collapse imphléascadh *m.*
collision imbhualadh *m.*
colloid crystal criostal *m.* collóide
colloid propulsion tiomáint *f.* chollóideach
colloidal dispersion system córas *m.* easraithe collóidigh

colloidal propellant tiomántán *m.* collóideach
colloidal system córas *m.* collóideach
color-magnitude diagram léaráid *f.* datha is méide
color dath *m.*
color blindness dathdhaille *f.*
color temperature dath-theocht *f.*
coma cóma *m.*
combined test comhthástáil *f.*
combustion dó *m.*; dóchan *m.*
combustion area achar *m.* dó
combustion chamber seomra *m.* dóchain; cuasán *m.* dó
combustion efficiency éifeachtacht *f.* dó
combustion experiment turgnamh *m.* dó
combustion instability éagobhsaíocht *f.* dó
combustion synthesis sintéis *f.* dó
comet cóiméad *m.*
command ordú *m.*
command antenna aeróg *f.* na n-orduithe
common bulkhead bulcaid *f.* chomónta
commonality[1] **(common or shared attribute)** tréith *f.* chomónta
commonality[2] **(the possession of like attributes or a like set of attributes)** comóntacht *f.*
communication cumarsáid *f.*
communication with extraterrestrial intelligence [CETI] cumarsáid *f.* le hintleacht eachtardhomhanda
communications satellite satailít *f.* chumarsáide
community reception glacadh *m.* pobail
co-moving coordinates comhordanáidí *fpl.* comhghluaisteacha
compact galaxy dlúthréaltra *m.*
companding compandáil *f.*; comhbhrú-fairsingiú *m.*
companion body cabhail *f.* chompánaigh
companion galaxy réaltra *m.* compánaigh
comparative planetology pláinéadeolaíocht *f.* chomparáideach
compatibility comhoiriúnacht *f.*
compatibility evaluation measúnú *m.* comhoiriúnachta
complex fluid sreabhán *m.* coimpléascach
compliance géilliúntas *m.*
component comhbhall *m.*

composite color image íomhá *f.* datha chumaisc
composite material ábhar *m.* ilchodach
composite propellant tiomántán *m.* ilchodach
compression comhbhrú *m.*
Compton effect éifeacht *f.* Compton
computer graphics ríomhghrafaic *f.*
computer-aided design [**CAD**] dearadh *m.* ríomhchuidithe [DRC]
computer-aided manufacturing [**CAM**] déantúsaíocht *f.* ríomhchuidithe; monarú *m.* ríomhchuidithe
computerized map mapa *m.* ríomhairithe
conceptual phase pas *m.* coincheapúil
condensation comhdhlúthú *m.*
configuration cumraíocht *f.*
configuration management bainistíocht *f.* cumraíochta
confined environment timpeallacht *f.* theoranta
confinement decondition díriochtú *m.* teorannaithe
conformal antenna aeróg *f.* chomhfhoirmiúil
conformance comhfhreagairt *f.*
conjunction cónasc *m.*
conquest concas *m.* (an spáis)
constant re-boost atreisiú *m.* tairiseach
constant-level balloon balún *m.* leibhéal-thairiseach
constellation réaltbhuíon *f.*
constitutional super-cooling sárfhuarú *m.* comhdhéanaimh
construction ratio cóimheas *m.* tógála
consultant sainchomhairleoir *m.*
container coimeádán *m.*
containerless processing próiseáil *f.* gan choimeádán
contamination éilliú *m.*; éilliúchán *m.*
continental shelf scairbh *f.* ilchríochach
contoured nozzle soc *m.* comhrianach
contract proposal togra *m.* conartha
contraction crapadh *m.*
contractor conraitheoir *m.*
contrast codarsnacht *f.*
contrast enhancement feabhsú *m.* codarsnachta
contravane contra-eite *f.*
contribution cion *m.*; alt *m.*

control rialú *m.*
control (to) rialú
control center lárionad *m.* rialaithe
control moment gyroscope [CMG] gíreascóp *m.* móiminte rialaithe
control of outer space rialú *m.* an spáis amuigh
control over space objects rialú *m.* réad spáis
control surface dromchla *m.* rialaithe
control vane eite *f.* rialaithe
controlled ecological life support system [CELSS] córas *m.* tacaíochta beatha éiceolaíche rialaithe; éiceachóras *m.* tacaíochta beatha rialaithe
convection-controlled comhiompar-rialaithe
convergent coinbhéirseach
convergent-divergent nozzle soc *m.* coinbhéirseach-eisréimneach
convolutional coding códú *m.* conbhlóideach
coolant fuarthán *m.*
cooling fuarú *m.*
coordination of space programs comhordú *m.* na gclár spáis
co-polar pattern patrún *m.* comhpholach
cord corda *m.*
core croí *m.*
Coriolis correction ceartú *m.* Coriolis
Coriolis force fórsa *m.* Coriolis
Coriolis parameter paraiméadar *m.* Coriolis
corona coróin *f.*
corona discharge díluchtú *m.* corónach
coronal heating téamh *m.* corónach
coronal mass ejection [CME] eisiachtain *f.* maise corónaí
corporation corparáid *f.*
corrective maintenance cotabháil *f.* cheartaitheach
correlation comhghaolú *m.*
cosmic background radiation rádaíocht *f.* chúlra chosmaigh; radaíocht *f.* chúlrach chosmach
cosmic dust deannach *m.* cosmach
cosmic infrared background (radiation) cúlra *m.* infridhearg cosmach (radaíocht)
cosmic law dlí *m.* cosmach
cosmic microwave background (radiation) cúlra *m.* micreathonnach cosmach (radaíocht)

cosmic model samhail *f.* chosmach
cosmic noise torann *m.* cosmach
cosmic power[1] (**country**) cumhacht *f.* chosmach
cosmic power[2] (**energy**) fuinneamh *m.* cosmach
cosmic radiation radaíocht *f.* chosmach
cosmic rays gathanna *mpl.* cosmacha
cosmic reionization athianú *m.* cosmach
cosmic space spás *m.* cosmach
cosmic time am *m.* cosmach
cosmic velocity treoluas *m.* cosmach
cosmic X-ray background cúlra *m.* X-gha cosmach
cosmochemistry cosmaicheimic *f.*
cosmochronology cosmachroineolaíocht *f.*
cosmodrome cosmadróm *m.*
cosmological constant tairiseach *m.* cosmeolaíoch
cosmological parameters paraiméadair *mpl.* chosmeolaíocha
cosmological principle prionsabal *m.* cosmeolaíoch
cosmological term téarma *m.* cosmeolaíoch
cosmological time am *m.* cosmeolaíoch
cosmology cosmeolaíocht *f.*
cosmonaut spásaire *m.*
cosmos cosmas *m.*
cost-plus-fee contract [**CPFF**] conradh *m.* le costas móide táille
cost-plus-fixed-fee contract [**CPFF**] conradh *m.* le costas móide táille shocraithe
cost-reimbursement contract conradh *m.* le costas aisíoctha
Couette flow sreabhadh *m.* Couette; sreabh *f.* Couette
countdown comhaireamh *m.* síos
countdown work obair *f.* comhaireamh síos
counter-glow frithloinnir *f.*
counter-measures frithbhearta *mpl.*
coupled modes móid *mpl.* chúplaithe
coverage són *m.* clúdaigh
crab crab *m.*; ionramháil *f.* phortáin
crab angle uillinn *f.* phortáin; uillinn *f.* chrab
cradle cliabhán *m.*
crew foireann *f.*
crew escape module modúl *m.* éalaithe foirne

crew time am *m.* foirne
critical cooling rate ráta *m.* criticiúil fuaraithe
critical design review [CDR] athbhreithniú *m.* criticiúil ar an dearadh; athbhreithniú *m.* dearaidh chriticiúil
critical Marangoni number uimhir *f.* Marangoni chriticiúil
critical phenomena feiniméin *mpl.* chriticiúla
criticality criticiúlacht *f.*
cross-fin traseite *f.*
cross-range trasraon *m.*
cross-ply laminate lannán *m.* crosdualach
cross-polar pattern patrún *m.* traspholach
crossflow cros-sreabhadh *m.*
crosstalk traschaint *f.*
crosswind trasghaoth *f.*
cryogen crióigin *f.*
cryogenic crióigineach
cryogenic arm géag *f.* chrióigineach
cryogenic propellant tiomántán *m.* crióigineach
cryogenic temperature teocht *f.* chrióigineach
cryogenics crióiginic *f.*
cryopumping criópumpáil *f.*
cryostat crióstat *m.*
crystal growth criostalfhás *m.*
crystallization criostalú *m.*; criostalúchán *m.*
culmination buaic *f.*
cumulative operating time am *m.* oibriúcháin carnach
curtailed inspection iniúchadh *m.* ciorraithe
cut-off scoitheadh *m.*
cylinder sorcóir *m.*

D

daily landing balloon balún *m.* tuirlingthe laethúla
damage damáiste *m.*
dark cloud scamall *m.* dorcha
dark energy fuinneamh *m.* dorcha
dark matter damhna *m.* dorcha
dark signal comhartha *m.* dorchadais
data sonraí *mpl.*
data acquisition fáil *f.* sonraí
data bus bus *m.* sonraí
data capture gabháil *f.* sonraí
data collection platform [**DCP**] ardán *m.* bailithe sonraí
data collection satellite [**DCS**] satailít *f.* bailithe sonraí
data entry iontráil *f.* sonraí
data processing próiseáil *f.* sonraí
data relay satellite [**DRS**] satailít *f.* athsheachadta sonraí
day lá *m.*
dayside (of a planet) taobh *m.* lae (de phláinéad)
dayside cusp[1] **(polar)** cuspa *m.* taobh lae; cuspa *m.* polach
dayside cusp[2] **(lunar)** rinn *f.*
de Laval nozzle soc *m.* de Laval
debrief (to) dífhaisnéis
debugging dífhabhtúchán *m.*; dífhabhtú *m.*
debugging test tástáil *f.* dífhabhtúcháin
decalcification díchailciú *m.*
deceleration luasmhoilliú *m.*
deceleration parachute paraisiút *m.* luasmhoillithe
declination diallas *m.*
decoder díchódóir *m.*
decoding díchódú *m.*
decompression dí-chomhbhrú *m.*
decompression chamber seomra *m.* díbhrúcháin; cuasán *m.* díbhrúcháin

decompression sickness tinneas *m.* dí-chomhbhrú
deconditioning díriochtú *m.*
decontamination dí-éilliú *m.*; dí-éilliúchán *m.*
decoupling díchúpláil *f.*
dedicated tiomnaithe
dedicated control center lárionad *m.* rialaithe tiomnaithe
de-emphasis díthreise *f.*
deep space cianspás *m.*
defect fabht *m.*
deflagration grodloscadh *m.*
deflation dítheannadh *m.*
degassing díghásáil *f.*
delimitation of airspace and outer space teorannú *m.* aerspáis agus an spáis amuigh
deluge collection pond linn *f.* bailithe díleann
demixing dímheascadh *m.*
dendrite deindrít *f.*
density-sliced image íomhá *f.* dhlús-slisnithe
de-orbit (**to**) dí-fhithisiú
de-orbiting dí-fhithisiú *m.*
dependent position determination cinneadh *m.* suímh spleách
depletion layer sraith *f.* ídíthe
depletion sensor braiteoir *m.* ídithe
deployable antenna aeróg *f.* inchurtha in úsáid
deployable structure struchtúr *m.* inchurtha in úsáid
depression lagrach *m.*; ísleán *m.*
depth of discharge [**DOD**] doimhneacht *f.* díluchtaithe
derating dírátú *m.*
descending node nód *m.* íslitheach
descent tuirlingt *f.*
descrambler díscrobhóir *m.*
design dearadh *m.*
design and development phase C pas *m.* dearaidh agus forbartha C
design authority údarás *m.* dearaidh
design concept phase pas *m.* choincheap dearaidh
design to cost [**DTC**] dearadh *m.* chun costais
design to lifecycle cost [**DTLCC**] dearadh *m.* chun costais saolré
designer dearthóir *m.*

desorption dí-asú *m.*; dí-asúchán *m.*

despin díchasadh *m.*

despin (**to**) díchasadh

despun díchasta

despun antenna aeróg *f.* dhíchasta

destination ceann *m.* scríbe

destruction of space objects scrios *m.* réad spáis

detached growth fás *m.* scortha

detailed design phase C pas *m.* miondearaidh C

detection of launches brath *m.* lainseálacha

detection threshold tairseach *f.* bhratha

detector brathadóir *m.*

detonating cord corda *m.* maidhmitheach

detonation madhmadh *m.*

detonator maidhmitheoir *m.*

development forbairt *f.*

development flight eitilt *f.* forbartha

development test tástáil *f.* forbartha

device gléas *m.*; feiste *f.*

Dewar fleascán *m.* Dewar

diaphragm spiara *m.*

differential manometer manaiméadar *m.* difreálach

differential pressure brú *m.* difreálach

differential pulse code modulation [**DPCM**] modhnú *m.* difreálach bíogchód

diffracted wave tonn *f.* dhíraonta

diffraction díraonachán *m.*; díraonadh *m.*

diffraction grating gríl *f.* díraonta

diffuse interstellar band banda *m.* idir-réaltach idirleata

diffuse reflection frithchaitheamh *m.* idirleata

diffuse reflector frithchaiteoir *m.* idirleata

diffuse sky radiation radaíocht *f.* spéire idirleata

diffusion coefficient comhéifeacht *f.* idirleata

diffusion-controlled idirleathadh-rialaithe

diffusion-limited idirleathadh-teoranta

digital digiteach

dilatable balloon balún *m.* inleata

dimensioning toisiú *m.*; toisiúchán *m.*

dipole antenna aeróg *f.* dhépholach
direct access dír-rochtain *f.*
direct broadcast satellite satailít *f.* craolta dírigh
direct broadcasting by satellite [DBS] craoladh *m.* díreach satailíte
direct distribution dáileadh *m.* díreach
direct motion dírghluaisne *f.*
direct orbit fithis *f.* dhíreach
direct re-entry athiontráil *f.* dhíreach
directional antenna treo-aeróg *f.*
directional gyro gíreascóp *m.* treoch
directional solidification soladú *m.* treoch
dirigible balloon balún *m.* instiúrtha
discharge nozzle soc *m.* díluchtaithe
discovery fionnachtain *f.*
discrete component comhbhall *m.* scoite
discrimination idirdhealú *m.*
dislocation easáitiú *m.*
dispersed particle cáithnín *m.* easraithe
dispersion mixing meascadh *m.* easraithe
dissemination of scientific data scaipeadh *m.* sonraí eolaíochta
dissipative structure struchtúr *m.* íditheach
distance measure tomhas *m.* faid
distance modulus modal *m.* faid
disturbed orbit fithis *f.* shuaite
divergent nozzle soc *m.* eisréimneach
DNA single- and double-strand breaks bristeacha *mpl.* dual singil agus dúbailte AND (DNA)
docking nascadh *m.*
docking module modúl *m.* nasctha
docking unit aonad *m.* nasctha
dog-leg ionramháil *f.* cos con; ionramháil *f.* threoraithe chumachtaithe
dome cruinneachán *m.*
Doppler drift síobadh *m.* Doppler
Doppler effect éifeacht *f.* Doppler
Doppler navigation loingseoireacht *f.* Doppler
Doppler shift aistriú *m.* Doppler
dose-equivalent dáileogchoibhéiseach
dose meter méadar *m.* dáileoige; dáileogmhéadar *m.*

dose measurement; dosimetry tomhas *m.* dáileoige; dáileogmhéadracht *f.*
dosimetry telescope [**DOSTEL**] teileascóp *m.* dáileogmhéadrachta
double-base propellant tiomántán *m.* bun dúbáilte
double-expansion nozzle soc *m.* fairsingithe dúbáilte
double integrating gyro gíreascóp *m.* suimeála dúbailte
doublet antenna aeróg *f.* dhúibléideach
down-range station stáisiún *m.* íosraoin
downlink íosnasc *m.*
drag cúltarraingt *f.*
drag free saor ó chúltarraingt
drag parachute paraisiút *m.* cúltarraingthe
Drake's formula foirmle *f.* Drake
drift síobadh *m.*
drift angle uillinn *f.* síobtha; síobuillinn *f.*
drift orbit fithis *f.* síobtha; síobfhithis *f.*
drift period tréimhse *f.* shíobtha; síobthréimhse *f.*
drogue parachute paraisiút *m.* tracóige
drogue recovery athfháil *f.* tracóige
drop tower túr *m.* titime
dropsonde dropshondáil *f.*; sondáil *f.* ligthe; sondáil *f.* anuas
dry weight meáchan *m.* tirim
dual-fuel débhreosla *m.*
dual-spin déchasadh *m.*
dummy stage céim *f.* bhréige; céim *f.* dhumaí
duplexer aeróg *f.* dhéphléacsach
dust absorption ionsú *m.* deannaigh
dust scattering scaipeadh *m.* deannaigh
dwarf galaxy abhacréaltra *m.*
dwarf planet abhacphláinéad *m.*
dwarf star abhacréalta *f.*
dynamic flattening maolú *m.* dinimiciúil
dynamic pressure brú *m.* dinimiciúil
dynamic stability cobhsaíocht *f.* dhinimiciúil
dynamical time am *m.* dinimiciúil
dynamo action gníomhú *m.* dineamó
dynamo theory teoiric *f.* dhineamó
dyspepsia dispeipse *f.*
dyspnea dispnia *f.*

E

earphone cluasán *m.*
Earth An Domhan *m.*
Earth crust screamh *f.* an Domhain
Earth exploration satellite service seirbhís *f.* satailíte taiscéalaíochta an Domhain
earth grounding talmhú *m.*; talmhúchán *m.*
Earth mean orbital speed [**EMOS**] meánluas *m.* fithiseach an Domhain
Earth orbit[1] (**Earth's orbit**) fithis *f.* an Domhain
Earth orbit[2] (**orbit around the Earth**) fithis *f.* thart ar an Domhan
Earth satellite satailít *f.* Domhain
Earth station stáisiún *m.* Domhain
Earth-to-orbit Domhan-go-fithis
Earth-to-orbit shuttle tointeálaí *m.* Domhan-go-fithis
earthing talmhú *m.*; talmhúchán *m.*
earthlight domhanloinnir *f.*; domhansolas *m.*
earthquake crith *m.* talún
earthshine domhanloinnir *f.*
ebullism aereambalacht *f.*
eccentric anomaly aimhrialtacht *f.* éalárnach
eccentric orbit fithis *f.* éalárnach
eccentricity éarlárnacht *f.*
eclipse urú *m.*; éiclips *m.*
eclipsing uraitheach
eclipsing binary dérealta *f.* uraitheach
eclipsing variable star réalta *f.* athraitheach uraitheach
ecliptic éiclipteach
ecliptic coordinate system córas *m.* comhordanáide éicliptí
ecological system córas *m.* éiceolaíoch
economic benefit sochar *m.* eacnamaíoch
Eddington limit teorainn *f.* Eddington
edge enhancement feabhsú *m.* ciumhaise
edge sharpening géarú *m.* ciumhaise

Edgeworth-Kuiper belt object [**EKBO**] réad *m.* creasa Edgeworth-Kuiper

effective control rialú *m.* éifeachtach

effective dose equivalent coibhéis *f.* dáileoige éifeachtaí

effective exhaust velocity treoluas *m.* sceite éifeachtach

effective isotropically radiated power [**EIRP**] cumhacht *f.* radaithe iseatrópach éifeachtach

effectiveness éifeachtacht *f.*

efflux deflector sraonaire *m.* eisilte

eight-bit byte beart *m.* ocht ngiotán

Einstein equation cothromóid *f.* Einstein

ejectable ineisteilgthe

ejectable fairing slíomú *m.* ineisteilgthe

ejection seat suíochán *m.* teilgin

ejector[1] (**one that ejects**) eisteilgeoir *m.*

ejector[2] (**ejecting device**) gléas *m.* eisteilgin; feiste *f.* eisteilgin

Ekman layer sraith *f.* Ekman

elastomer leaistiméir *m.*

electric discharge díluchtú *m.* leictreach

electric power converter [**EPC**] tiontaire *m.* cumhachta leictreach

electric propulsion tiomáint *f.* leictreach

electro-dynamic vibration generator gineadóir *m.* creathaidh leictridinimiciúil

electro-dynamics leictridinimic *f.*

electro-jet leictreascaird *f.*

electromagnetic compatibility [**EMC**] comhoiriúnacht *f.* leictreamaighnéadach

electromagnetic disturbance suaitheadh *m.* leictreamaighnéadach

electromagnetic gun gunna *m.* leictreamaighnéadach

electromagnetic interference [**EMI**] trasnaíocht *f.* leictreamaighnéadach

electromagnetic launcher lainseálaí *m.* leictreamaighnéadach

electromagnetic levitation eadarbhuasú *m.* leictreamaighnéadach

electromagnetic pulse [**EMP**] bíog *f.* leictreamaighnéadach

electromagnetic remote sensing cianbhraiteacht *f.* leictreamaighnéadach

electromyogram leictrimíogram *m.*

electronic counter-measure frithbheart *m.* leictreonach

electrophoresis leictreafóiréis *f.*
electrophoresis method modh *m.* leictreafóiréise
electrostatic levitation eadarbhuasú *m.* leictreastatach
electrostatic propulsion tiomáint *f.* leictreastatach
electrothermal hydrazine thruster ceartaitheoir *m.* hiodraisín leictriteirmeach
electrothermal propulsion tiomáint *f.* leictriteirmeach
electrothermal thruster ceartaitheoir *m.* leictriteirmeach
elevon airdearán *m.*
elliptical galaxy réaltra *m.* éilipseach
ellipticity éilipseacht *f.*
elongation fadú *m.*; fadúchán *m.*
emergency éigeandáil *f.*
emerging magnetic flux flosc *m.* maighnéadach éiritheach; flosc *m.* maighnéadach atá ag teacht chun cinn
emission line líne *f.* astaithe
emissivity astaíochas *m.*
emittance astaíocht *f.*
empennage iarchuid *f.*
emphysema eimfiséime *f.*
encoder ionchódóir *m.*
encoding ionchódú *m.*
encounter teagmháil *f.*
end of life [EOL] deireadh *m.* ré
end-of-life phase pas *m.* deireadh ré
endogenous sources of organic molecules foinsí *fpl.* inginiúla na móilíní orgánacha
endurance test tástáil *f.* fulaingthe
energetic particles cáithníní *mpl.* fuinneamhacha
energy density dlús *m.* fuinnimh
energy physics fisic *f.* fuinnimh
energy volume toirt *f.* fuinnimh
engine inneall *m.*
engine frame fráma *m.* innill
engine manufacturer déantúsóir *m.* innill
engineering model [EM] samhail *f.* innealtóireachta
entry[1] **(atmospheric)** iontráil *f.*
entry[2] **(planetary)** iontráil *f.*

envelope imchlúdach *m.*
environment timpeallacht *f.*
environmental monitoring monatóireacht *f.* timpeallachta
environmental test tástáil *f.* timpeallachta
ephemeris eifirmiris *f.*
equatorial coordinates comhordanáidí *fpl.* meánchriosacha
equatorial orbit fithis *f.* mheánchriosach
equatorial plane plána *m.* meánchriosach
equilibrium cothromaíocht *f.*
equilibrium (theory of) tide taoide *f.* chothromaíochta (teoiric chothromaíochta taoidí)
equilibrium disturbance suaitheadh *m.* cothromaíochta
equilibrium mixture meascán *m.* cothromaíochta
equinoctial point pointe *m.* cónochta
equinoctial system of coordinates córas *m.* cónochta comhordanáidí
equinox cónocht *m.*
equipment trealamh *m.*
equipment manufacturer déantúsóir *m.* trealaimh
equivalent isotropic radiated power [EIRP] cumhacht *f.* radaithe iseatrópach choibhéiseach
erytheme eiritéime *f.*
erythrocyte eiritricít *f.*
escape tower túr *m.* éalaithe
escape velocity treoluas *m.* éalaithe
Euler period tréimhse *f.* Euler
Eulerian angles uillinneacha *fpl.* Euler
Eulerian coordinates comhordanáidí *fpl.* Euler
Eulerian equations cothromóidí *fpl.* Euler
Europa Europa *f.*
evaluation test tástáil *f.* luachála
evaporation rate ráta *m.* galaithe
event camera ceamara *m.* teagmhais
event controller rialaitheoir *m.* teagmhais
evidence fianaise *f.*
exclusive right ceart *m.* eisiach
exhaust sceitheadh *m.*
exhaust velocity treoluas *m.* sceite
exobase eiseabhun *m.*

exobiology eisibhitheolaíocht *f.*
exocentric angle uillinn *f.* eisealárnach
exogenous delivery of organic molecules seachadadh *m.* eisginiúil na móilíní orgánacha
exogenous sources of organic molecules foinsí *fpl.* eisginiúla na móilíní orgánacha
exoplanet eiseaphláinéad *m.*
exosphere eisisféar *m.*
expanding balloon balún *m.* fairsingithe; balún *m.* fairsingíoch
expansion fairsingiú *m.*
expendable inídithe
expendable launch vehicle [**ELV**] feithicil *f.* lainseála inídithe
experience taithí *f.*; cleachtadh *m.*
experiment turgnamh *m.*
experiment package pacáiste *m.* turgnaimh
expert appraisal measúnacht *f.* shaineolach
expert system córas *m.* saineolach
exploding bridge wire [**EBW**] sreang *f.* phléascach droichid
exploding wire sreang *f.* phléascach
exploitation of resources saothrú *m.* acmhainní
explosion pléascadh *m.*
explosive bolt bolta *m.* pléascach
explosive train sraith *f.* phléascach
exposure nochtadh *m.*
exposure facility saoráid *f.* nochta
extended red emission deargastú *m.* sínte
external burning dó *m.* seachtrach
external magnetism maighnéadas *m.* seachtrach
external magnetosphere maighnéadaisféar *m.* seachtrach
extinct nuclide núiclíd *f.* dhíofa; núiclíd *f.* mharbh
extinction díobhadh *m.*
extinction phenomenon feiniméan *m.* díobhaidh
extrados easdroim *m.*
extrasolar planet pláinéad *m.* eachtarghréine
extraterrestrial eachtardhomhanda
extraterrestrial environment timpeallacht *f.* eachtardhomhanda
extraterrestrial intelligence [**ETI**] intleacht *f.* eachtardhomhanda
extraterrestrial life beatha *f.* eachtardhomhanda

extravehicular activity [EVA] gníomhaíocht *f.* seachfheithicile
extravehicular mobility unit [EMU] aonad *m.* soghluaiseachta seachfheithicile
extreme environments timpeallachtaí *fpl.* foircneacha
extreme life beatha *f.* fhoircneach
extremophiles foircnifiligh *mpl.*
eye flash splanc *f.* shúile
eye-head coordination comhordú *m.* súile agus cinn

F

factor fachtóir *m.*

facula (pl. faculae) facalach *m.*

fade-out céimniú *m.* as

fading céimniú *m.*

fail-safe slán ó theip

failure teip *f.*; cliseadh *m.*

failure intensity déine *f.* teipe

failure modes and effects analysis [**FMEA**] anailís *f.* mód teipe agus éifeachtaí

failure modes effects and criticality analysis [**FMECA**] anailís *f.* mód teipe éifeachtaí agus criticiúlachta

failure rate ráta *m.* teipe(anna)

faint (**celestial**) **object** réad *m.* (neamhaí) fann

fairing slíomú *m.*

false color dath *m.* bréige

fan-beam antenna aeróg *f.* léas gaothráin

Faraday cage cás *m.* Faraday

Faraday rotation rothlú *m.* Faraday

farside of the Moon an taobh *m.* thall den Ghealach

fascination meallacacht *f.*

fast Fourier transform [**FFT**] trasfhoirm *f.* thapaidh Fourier

fast-twitch muscles matáin *mpl.* mhearbhíogacha

fault locht *m.*; fabht *m.*

feasibility indéantacht *f.*

feasibility assessment phase pas *m.* measúnaithe indéantachta

feasibility study staidéar *m.* indéantachta

feasibility study phase pas *m.* staidéir indéantachta

feed fotha *m.*; fothú *m.*

feedback aischothú *m.*; aiseolas *m.*; aisfhotha *m.*

feeder (in heat pipe) cothaitheoir *m.* (i bpíobán teasa)

feeder link nasc *m.* cothaitheora

feed-line líne *f.* fotha

feed-system córas *m.* fotha
feed-forward réamhchothú *m.*; tulchothú *m.*; tulfhotha *m.*
fibre optics snáthoptaic *f.*
field data sonraí *mpl.* allamuigh
field effect transistor [FET] trasraitheoir *m.* tionchar réimse
field maintenance work obair *f.* chotabhála allamuigh
field of view réimse *m.* radhairc
filament filiméad *m.*
filament winding tochrán *m.* filiméid
film cooling fuarú *m.* scannáin
fin eite *f.*
firing room seomra *m.* scaoilte; seomra *m.* lainseála
firm fixed price praghas *m.* socraithe daingean
firm fixed-price contract conradh *m.* ar phraghas socraithe daingean
first cosmic velocity an chéad treoluas *m.* cosmach
first stage an chéad chéim *f.*
fishbone antenna aeróg *f.* cnámh éisc
fit oiriúnach
fixed price praghas *m.* socraithe
fixed-price contract conradh *m.* ar phraghas socraithe
fixed-price incentive contract conradh *m.* le dreasacht praghais socraithe
fixed-price with price variation contract conradh *m.* ar phraghas socraithe le hathrú praghais
fixed satellite service seirbhís *f.* satailíte socraithe
fixed unit price praghas *m.* aonaid socraithe
flame chute siúit *f.* lasrach
flame deflector sraonaire *m.* lasrach
flameout múchadh *m.* lasrach
flap plapa *m.*
flare bladhm *f.*; bladhaire *m.*
flare star bladhmréalta *f.*
flashback cúlsplanc *f.*
flat spin réchasadh *m.*
flat universe cruinne *f.* chomhréidh
flatness problem fadhb *f.* chomhréidhe
flexible linear shaped charge lucht *m.* línechruthach solúbtha
flight eitilt *f.*
flight commands orduithe *mpl.* eitilte

flight control rialú *m.* eitilte
flight experiment turgnamh *m.* eitilte
flight model [**FM**] samhail *f.* eitilte
flight readiness review [**FRR**] athbhreithniú *m.* ullmhachta eitilte
flight sequence seicheamh *m.* eitilte
flight sequence check examination scrúdú *m.* seiceála seicheamh eitilte
flight test tástáil *f.* eitilte
flight train triopall *m.* eitilte
flight unit aonad *m.* eitilte
float altitude airde *f.* foluana
floating-zone method [**FZ method**] modh *m.* snámhshón (modh *m.* FZ)
flowchart sreabhchairt *f.*
fluid instability éagobhsaíocht *f.* sreabháin
fluid physics experiment turgnamh *m.* fisice sreabhán
fluid shift aistriú *m.* sreabháin
flush-mounted antenna aeróg *f.* réshuite
flutter eitleach *m.*
fluxmeter floscmhéadar *m.*
fly-by neaseitilt *f.*
foamed plastics cúrphlaisticí *fpl.*
folded dipole antenna aeróg *f.* dhépholach fhillte
folding fin eite *f.* fhillteach
fold-over distortion díchumadh *m.* aisfhillteach
footprint of a beam lorg *m.* léis
forbidden line líne *f.* thoirmeasctha
Forbush decrease laghdú *m.* Forbush
forced convection comhiompar *m.* fórsáilte
foreshock réamhchrith *m.*
forming foirmiú *m.*
forward contamination tuléilliú *m.*
forward contamination prevention cosc *m.* ar thuléilliú
forward dome tulchruinneachán *m.*
forward scattering tulscaipeadh *m.*
forward test post [**FTP**] tulphost *m.* rialaithe
frame fráma *m.*
Fraunhofer lines línte *fpl.* Fraunhofer
free access saor-rochtain *f.*
free atmosphere saor-atmaisféar *m.*

free balloon saorbhalún *m.*
freefall saorthitim *f.*
free-fall tower túr *m.* saorthitime
freeflow electrophoresis leictreafóiréis *f.* shaorshreafa
free-flying satellite satailít *f.* saoreitilte
free growth saorfhás *m.*
free interface saor-idiréadan *m.*
free lift saorardú *m.*
free-stream temperature teocht *f.* saorshrutha
free surface saordhromchla *m.*
frequency allocation leithdháileadh *m.* minicíochta
frequency channel cainéal *m.* minicíochta
frequency-division multiple access [**FDMA**] ilrochtain *f.* roinnte minicíochta
frequency-division multiplexing [**FDM**] ilphléacsú *m.* roinnte minicíochta
frequency modulation [**FM**] modhnú *m.* minicíochta
frequency re-use athúsáid *f.* minicíochta
frequency re-used satellite network líonra *m.* satailíte le hathúsáid minicíochta
frequency shift keying [**FSK**] eochrú *m.* iomlaoid mhinicíochta
frozen-in field réimse *m.* reoite isteach
frozen mixture meascán *m.* reoite
frustration frustrachas *m.*
frustration threshold tairseach *f.* frustrachais
fuel breosla *m.*
fuel cell breoslachill *f.*
fuel nozzle soc *m.* breosla
fuel shut-off stopadh *m.* breosla
fuel system córas *m.* breosla
fuel-to-oxidizer ratio cóimheas *m.* breosla go hocsaídeoir
fueling vehicle feithicil *f.* bhreosla
fuelman fear *m.* breosla
full pressure-suit lán-bhrúchulaith *f.*
functional residual capacity [**FRC**] toilleadh *m.* iarmharach feidhme
fundamental mode of vibration mód *m.* creathaidh bunúsach
fundamental plane plána *m.* bunúsach
fuse cord corda *m.* fiúis
fusion reaction imoibriú *m.* comhleá

G

gain over noise temperature [G/T] neartú *m.* thar theocht torainn
gaiter loirgneán *m.* deiridh roicéid
gal gal *m.*
galactic bulge bolg *m.* réaltrach
galactic center lár *m.* réaltrach
galactic chimney cuas *m.* réaltrach
galactic cloud néal *m.* réaltrach; scamall *m.* réaltrach
galactic collapse imphléascadh *m.* réaltrach
galactic coordinates comhordanáidí *fpl.* réaltracha
galactic cosmic radiation [GCR] radaíocht *f.* chosmach réaltrach
galactic cosmic ray ga *m.* cosmach réaltrach
galactic disk diosca *m.* réaltrach
galactic fountain scairdeán *m.* réaltrach
galactic halo luan *m.* réaltrach
galactic magnetic field réimse *m.* maighnéadach réaltrach
galactic plane plána *m.* réaltrach
galactic ridge X-ray emission astú *m.* X-ghathach droma réaltraigh
galactic wind gaoth *f.* réaltrach
galaxy réaltra *m.*
Galaxy An Réaltra *m.*; Bealach *m.* na Bó Finne
galaxy distribution scaipeadh *m.* réaltra(í)
galaxy evolution model samhail *f.* éabhlóide réaltra
gamma astronomy réalteolaíocht *f.* ghámа
gamma burst rois *f.* gháma
gamma-ray burst rois *f.* gháma -ghathach
gantry túr *m.* freastail
gantry crane craein *f.* droichid
garbage dramhaíl *f.*
gas atmosphere atmaisféar *m.* gáis
gas generator gineadóir *m.* gáis
gas vane eite *f.* gháis
gas-buffered heat pipe píobán *m.* teasa gásmhaolánach

gas-liquid separation deighilt *f.* gháis is leachta
gas-turbine engine inneall *m.* gástuirbín
gegenschein frithloinnir *f.*
genetic effects of radiation éifeachtaí *fpl.* géiniteacha radaíochta
geocentric angle uillinn *f.* gheolárnach
geocentric longitude fad *m.* geolárnach
geocentric zenith forar *m.* geolárnach
geodesy geodasaíocht *f.*
geodetic satellite satailít *f.* gheodasach
geodetic zenith forar *m.* geodasach
geodynamics geoidinimic *f.*
geographic longitude fad *m.* geografach
geoid geoid *f.*
geomagnetic field réimse *m.* geomaighnéadach; réimse *m.* maighnéadach an Domhain
geomagnetic storm stoirm *f.* gheomaighnéadach
geomagnetic tail eireaball *m.* geomaighnéadach
geomagnetism geomaighnéadas *m.*
geometric albedo ailbéideacht *f.* gheoiméadrach
geometric flattening leacú *m.* geoiméadrach
geophysics geoifisic *f.*
geopotential geopoitéansal *m.*
geostationary Earth orbit [**GEO**] fithis *f.* Domhain gheochobhsaí
geostationary satellite satailít *f.* gheochobhsaí
geostationary satellite orbit [**GSO**] fithis *f.* satailíte geochobhsaí
geostationary transfer orbit [**GTO**] fithis *f.* aistrithe gheochobhsaí
geostrophic wind gaoth *f.* gheostrófach
geosynchronous satellite satailít *f.* gheoisioncronach
geotropism geotrópacht *f.*
getter ábhar *m.* súite
getting sú *m.*
g-force g-fhórsa *m.*; fórsa *m.* imtharraingthe
giant impact theory teoiric *f.* tuinsimh mhóir
giant star fathachréalta *f.*
gigayear gigibhliain *f.*
gimbal (to) giombalú
g-jitter g-ghiodam *m.*
glide foluain *f.*

glider faoileoir *m.*
global domhanda
global change athrú *m.* domhanda
globular cluster réaltbhraisle *f.* chruinneogach; braisle *f.* chruinneogach
glovebox lámhainnbhosca *m.*
gondola gandala *m.*
governmental control rialú *m.* rialtais
gradient wind gaoth *f.* ghrádáin
gradiometer grádaiméadar *m.*
grain gráinne *m.*
granulation gránú *m.*; granúchán *m.*
granule gráinnín *m.*
granulometry gráinnínmhéadracht *f.*
grating gríl *f.*
graveyard orbit fithis *f.* reilige; fithis *f.* deiridh satailíte
gravikinesis meá-chinéis *f.*
gravimetric energy fuinneamh *m.* meá-mhéadrach
gravimorphism meá-mhorfacht *f.*
gravimorphogenesis meá-mhorfaigineas *m.*
graviperception meá-aireachtáil *f.*
gravireceptor meá-ghabhdóir *m.*
gravireceptors meá-ghabhdóirí *mpl.*
graviresponse meá-fhreagairt *f.*
gravisensation meá-mhothú *m.*; meá-mhothúchán *m.* mothú *m.* imtharraingthe
gravisensing meá-bhrath *m.*
gravisensitivity meá-íogaireacht *f.*
gravisensor meá-bhraiteoir *m.*
gravistimulation meá-bhrostú *m.*
gravitation imtharraingt *f.*
gravitation stress strus *m.* imtharraingteach
gravitational imtharraingteach
gravitational biology bitheolaíocht *f.* imtharraingteach
gravitational constant tairiseach *m.* na himtharraingthe
gravitational contraction crapadh *m.* imtharraingteach
gravitational field réimse *m.* imtharraingteach
gravitational instability éagobhsaíocht *f.* imtharraingteach
gravitational lens lionsa *m.* imtharraingteach

gravitational loading lódáil *f.* imtharraingteach
gravitational many-body problem fadhb *f.* ilchorp imtharraingteach
gravitational potential poitéansal *m.* imtharraingteach
gravitational red shift deargaistriú *m.* imtharraingteach
gravitational response freagairt *f.* imtharraingteach
gravitational segregation leithscaradh *m.* imtharraingteach
gravitational unloading dílódáil *f.* imtharraingteach
gravitational wave tonn *f.* imtharraingteach
gravitaxis meá-luadar *m.*
gravitropism meá-thrópacht *f.*
gravity imtharraingt *f.*; domhantarraingt *f.*
gravity convection comhiompar *m.* imtharraingthe
gravity fluctuation iomlaoid *f.* imtharraingthe
gravity gradient grádán *m.* imtharraingthe
gravity-gradient stabilization cobhsaíocht *f.* le grádán imtharraingthe
gravity gradiometer grádaiméadar *m.* imtharraingthe; grádaiméadar *m.* domhantarraingthe
gravity sensor braiteoir *m.* imtharraingthe
gravity threshold tairseach *f.* imtharraingthe
Great Dark Spot An Spota *m.* Mór Dorcha
Great Red Spot An Deargspota *m.* Mór
Great White Spot An Bánspota *m.* Mór
greenhouse effect iarmhairt *f.* cheaptha teasa
Greenwich mean time [GMT] meán-am *m.* Greenwich [MAG]
Gregorian calendar féilire *m.* Ghréagóra
Gregorian system córas *m.* Gréagórach
grooved heat pipe píobán *m.* teasa eitreach
ground talamh *f.*
ground control group grúpa *m.* rialaithe talún
ground patch area achar *m.* geadáin talún
ground segment deighleog *f.* talún
ground station stáisiún *m.* talún
ground swath sraith *f.* talún
ground test tástáil *f.* talún
ground track rian *m.* talún
ground truth fírinne *f.* na talún
ground-based simulation insamhladh *m.* ón talamh
grounding[I] <US> (earthing) talmhú *m.*; talmhúchán *m.*

grounding[2] <US> **(prohibit from flying)** bac *m.* ar eitilt
group of galaxies grúpa *m.* réaltraí
g-suit g-chulaith *f.*
g-tolerance g-fhulaingt *f.*; fulaingt *f.* imtharraingthe
guidance treoir *f.*; treorú *m.*; treoraíocht *f.*
guided missile diúracán *m.* treoraithe
guiderope rópa *m.* treorach
Gunn-Peterson effect éifeacht *f.* Gunn-Peterson
gyration gíoráil *f.*
gyration period tréimhse *f.* ghíorála
gyro-frequency gírimhinicíocht *f.*
gyro-horizon gíriléaslíne *f.*
gyro-laser gíriléasar *m.*
gyro-rotor gírearótar *m.*
gyro-stabilized platform ardán *m.* gíreachobhsaithe
gyro-unit gírea-aonad *m.*
gyro-wheel gírearoth *m.*

H

habitability ináitritheacht *f.*
habitable zone són *m.* ináitrithe
habitation module modúl *m.* áitribh
haemodynamics haemidinimic *f.*
hair cell cill *f.* ghruaige
Hall effect éifeacht *f.* Hall
hard copy crua-chóip *f.*
hard landing crua-thuirlingt *f.*
hardening cruachan *f.*
hardware crua-earraí *mpl.*
Hawking radiation radaíocht *f.* Hawking
Hayashi track conair *f.* Hayashi
hazard guais *f.*
haze smúiteán *m.*
headache tinneas *m.* cinn
head-down display [**HDD**] taispeáint *f.* chromchinn; taispeáint *f.* ceann síos
head-down tilt claonadh *m.* ceann síos
head-down tilt bed rest taca *m.* droma claonta ceann síos
head-to-foot acceleration luasghéarú *m.* ceann go cos
head-up display [**HUD**] taispeáint *f.* cheannard; taispeáint *f.* ceann suas
head-up tilt [**HUT**] claonadh *m.* ceann suas
heat balance cothromaíocht *f.* teasa
heat barrier bacainn *f.* teasa
heat blockage blocáil *f.* teasa
heat of ablation teas *m.* eisídithe
heat of fusion teas *m.* comhleá
heat of sublimation teas *m.* néalúcháin
heat of vaporization teas *m.* galaithe
heat pipe píobán *m.* teasa
heat-pipe radiator radaitheoir *m.* píobáin teasa
heat screen scáileán *m.* teasa

heat shield sciath *f.* theasa; teas-sciath *f.*
heat sink slogaide *f.* teasa
heavenly body rinn *m.* neimhe
heavy-case test tástáil *f.* i gcás crua
heavy ion ian *m.* trom
heavy-particle beam léas *m.* cáithnín throim
heliocentric longitude fad *m.* héilealárnach
heliocentric theory teoiric *f.* héilealárnach
heliopause héileastad *m.*
helioseismology héiliseismeolaíocht *f.*
heliosphere héilisféar *m.*
Helmholtz instability éagobhsaíocht *f.* Helmholtz
Herbig-Haro object réad *m.* Herbig-Haro
Hertzsprung-Russell diagram léaráid *f.* Hertzsprung-Russell
heterogeneous nucleation núicléatú *m.* ilchineálach
heterogeneous propellant tiomántán *m.* ilchineálach
heterosphere heitrisféar *m.*
Higgs field réimse *m.* Higgs
high-altitude balloon balún *m.* airde mhór
high-altitude flight eitilt *f.* airde mhór
high-altitude nuclear test tástáil *f.* núicléach airde mhór
high-definition television [HDTV] [HDT] teilifís *f.* ardghléine
high-energy cosmic radiation radaíocht *f.* chosmach ardfhuinnimh
high-energy heavy particle tromcháithnín *m.* ardfhuinnimh
high-energy space radiation spásradaíocht *f.* ardfhuinnimh
high-level data link control [HDLC] rialú *m.* nasctha sonraí ardleibhéil
high-pass filter scagaire *m.* ardphasach
high-velocity star réalta *f.* ardluais
highly eccentric orbit fithis *f.* ard-éalárnach
HII region réigiún *m.* HII
Hilda group grúpa *m.* Hilda
Hill radius raidias *m.* Hill; ga *m.* Hill
hindlimb suspension crochadh *m.* géige deiridh
hinge moment móimint *f.* inse
Hohmann transfer aistriú *m.* Hohmann
homeostasis hoiméastáis *f.*
homogeneous dispersion easrú *m.* aonchineálach
homogeneous nucleation núicléatú *m.* aonchineálach

homogeneous propellant tiomántán *m.* aonchineálach
homosphere homaisféar *m.*
hop[1] **(short flight)** eitilt *f.* ghairid
hop[2] **(bound)** preab *f.*
horizon problem fadhb *f.* léaslíne
horizon system of coordinates córas *m.* léaslíne na gcomhordanáidí
horizontal branch star réalta *f.* brainse chothrománaigh
horizontal coordinates comhordanáidí *fpl.* cothrománacha
horizontal plane plána *m.* cothrománach
host computer óstríomhaire *m.*
hot gas gás *m.* te
hot-air balloon balún *m.* aeir the
hour angle uair-uillinn *f.*
hour circle uairchiorcal *m.*
housekeeping bainistíocht *f.*
housekeeping telemetry teiliméadracht *f.* bhainistíochta
Hubble classification rangú *m.* Hubble
Hubble constant tairiseach *m.* Hubble
Hubble Deep Field [**HDF**] cianréimse *m.* Hubble
Hubble diagram léaráid *f.* Hubble
Hubble flow sreabhadh *m.* Hubble; sreabh *f.* Hubble
Hubble's law dlí *m.* Hubble
human behavior iompar *m.* daonna
human factors tosca *fpl.* daonna
human factors engineering innealtóireacht *f.* tosca daonna
human phantom taibhse *f.* dhaonna
human space flight spáseitilt *f.* dhaonna
hybrid (integrated) circuit ciorcad *m.* (comhtháite) hibrideach
hybrid propellant tiomántán *m.* hibrideach
hybrid resonance athshondas *m.* hibrideach
hydraulic vibration generator gineadóir *m.* creathaidh hiodrálach
hydrometeor hidrimeitéar *m.*
hydrostatic pressure brú *m.* hidreastatach
hydrothermal vent gaothairí *mpl.* hidriteirmeacha
hyperballistic hipearbalaistíoch
hyperballistic tunnel tollán *m.* hipearbalaistíoch
hyperballistics hipearbalaistíocht *f.*
hyperbolic guidance treoraíocht *f.* hipearbóileach

hyperbolic navigation loingseoireacht *f.* hipearbóileach
hypergolic propellant tiomántán *m.* hipeargóileach
hypergravity hipir-imtharraingt *f.*
hypermetropia hipirmeatróipe *f.*
hyperopia hipearóipe *f.*
hyperphoria hipearfóiria *f.*
hypersonic hipearsonach
hypersonic flow sreabhadh *m.* hipearsonach; sreabh *f.* hipearsonach
hypersonics hipearsonaic *f.*
hyperthermophile hipirteirmifileach *m.*
hypocapnia hipeacaipnia *f.*
hypoglycemia hipiglicéime *f.*
hypogravity hip-imtharraingt *f.*
hypokinesy hipicinéis *f.*
hypotension fotheannas *m.*
HZE particle cáithnín *m.* HZE

I

identification aithint *f.*; aitheantas *m.*
igniter adhainteoir *m.*
ignition adhaint *f.*
ignition delay moill *f.* adhainte
illumination soilsiú *m.*
image enhancement feabhsú *m.* íomhá
image registration clárú *m.* íomhá
image restoration athchóiriú *m.* íomhá
image tube feadán *m.* íomhá
imagery íomháineachas *m.*
imaginary time am *m.* samhailteach
imaging íomháú *m.*
imaging device gléas *m.* íomháithe
imaging instrument ionstraim *f.* íomháithe
imaging radar radar *m.* íomháineachais
imaging spectrometer speictriméadar *m.* íomháithe
imaging system córas *m.* íomháithe
immune response freagairt *f.* imdhíonach
impact pressure brú *m.* tuinsimh
impact sterilization steiriliú *m.* tuinsimh
impact tube feadán *m.* tuinsimh
impeller brúiteoir *m.*
impeller vane eite *f.* bhrúiteora
implement (to) cur i ngníomh
implementation cur *m.* i ngníomh
in-situ observation breathnóireacht *f.* in situ; breathnú *m.* ar an láithreán
in-space propulsion tiomáint *m.* ionspáis
incident teagmhas *m.*
incoherent scattering scaipeadh *m.* neamhleanúnach
independent position determination cinneadh *m.* suímh neamhspleách
individual reception glacadh *m.* aonair
induced environment timpeallacht *f.* ionduchtaithe

inert támh
inertial guidance treoraíocht *f.* thámhúil
inertial navigation loingseoireacht *f.* thámhúil
inertial platform ardán *m.* támhúil
inertial reference frame fráma *m.* tagartha támhúil; tagairtfhráma *m.* támhúil
inertial reference system córas *m.* tagartha támhúil
inertial unit aonad *m.* támhúil
inferior conjunction cónasc *m.* íochtair
inflation tube feadán *m.* teannta
infrahuman infreadhaonna
infrared astronomy réalteolaíocht *f.* infridhearg
infrared galaxy réaltra *m.* infridhearg
infrared star réalta *f.* infridhearg
initial mass function feidhm *f.* maise tosaigh
initiation tionscaint *f.*; tionscnamh *m.*
injection instealladh *m.*
injector insteallaire *m.*
inner planet pláinéad *m.* laistigh
insolation grianrú *m.*
insomnia easuan *m.*
inspect (to) iniúchadh
inspection iniúchadh *m.*
instrument ionstraim *f.*
instrument landing system [ILS] córas *m.* tuirlingthe ionstraime
intake pressure brú *m.* iontógála
integral albedo ailbéideacht *f.* shuimeálach
integral tank umar *m.* suimeálach
integrated circuit [IC] ciorcad *m.* comhtháite
integrated energy fuinneamh *m.* comhtháite
integration suimeáil *f.*
integration model samhail *f.* suimeála
intelligence intleacht *f.*
interactive idirghníomhach
interactive combustion dó *m.* idirghníomhach
intercontinental ballistic missile diúracán *m.* balaistíoch idir-ilchríochach
interface comhéadan *m.*
interfacial tension teannas *m.* idiréadanach

interference trasnaíocht *f.*
interfering signal comhartha *m.* trasnaíochta
interfering source foinse *f.* trasnaíochta
interferometry trasnamhéadracht *f.*
intergalactic space spás *m.* idir-réaltrach
interior design dearadh *m.* intí
intermediate frequency [IF] minicíocht *f.* idirmheánach
intermediate mass black hole dúpholl *m.* maise idirmheánaí
internal burning dó *m.* inmheánach
internal magnetism maighnéadas *m.* inmheánach
internal magnetosphere maighnéadaisféar *m.* inmheánach
international atomic time [IAT] am *m.* adamhach idirnáisiúnta [AAI]
interorbital transfer aistriú *m.* idirfhithiseach
interplanetary communication cumarsáid *f.* idirphláinéadach
interplanetary dust deannach *m.* idirphláinéadach
interplanetary space spás *m.* idirphláinéadach
interruption idirscor *m.*; idirbhriseadh *m.*; briseadh *m.*
inter-satellite link nasc *m.* idirshatailíte
interstage bay bá *f.* idirchéime
interstage skirt sciorta *m.* idirchéime
interstage structure struchtúr *m.* idirchéime
interstellar cloud scamall *m.* idir-réaltach
interstellar dust deannach *m.* idir-réaltach
interstellar extinction díobhadh *m.* idir-réaltach
interstellar magnetic field réimse *m.* maighnéadach idir-réaltach
interstellar medium meán *m.* idir-réaltach
interstellar molecular clouds scamaill *mpl.* mhóilíneacha idir-réaltacha
interstellar molecule móilín *m.* idir-réaltach
interstellar polarization polarú *m.* idir-réaltach
interstellar reddening deargadh *m.* idir-réaltach
interstellar shock wave tonn *f.* turrainge idir-réaltach
interstellar space spás *m.* idir-réaltach
intersymbol interference trasnaíocht *f.* idirshiombail
intersystem compatibility comhoiriúnacht *f.* idirchóras
intrados iondroim *m.*
intrasystem compatibility comhoiriúnacht *f.* laistigh de chóras
investigation imscrúdú *m.*; iniúchadh *m.*; fiosrú *m.*
investigation test tástáil *f.* imscrúdaithe

ion gauge tomhsaire *m.* iain
ion propulsion tiomáint *f.* iain
ionization gauge tomhsaire *m.* ianaithe
ionized exobase eiseabhun *m.* ianaithe
ionized exosphere eisisféar *m.* ianaithe
ionizer ianaitheoir *m.*
ionized hydrogen region réigiún *m.* hidirigine ianaithe
ionizing radiation radaíocht *f.* ianaíoch
ion-molecular reaction imoibriú *m.* ianmhóilíneach
ionophore ianafór *m.*
ionosphere ianaisféar *m.*
ionospheric disturbance suaitheadh *m.* ianaisféarach
ionospheric propagation forleathadh *m.* ianaisféarach
ionospheric sounding sondáil *f.* ianaisféarach
ionospheric storm stoirm *f.* ianaisféarach
iron meteorite dreigít *f.* iarainn; dreigít *f.* fheiriúil
irradiance ionradantas *m.*
irradiation ionradaíocht *f.*
irregular galaxy réaltra *m.* neamhrialta
isophote iseafót *m.*
isotope anomaly aimhrialtacht *f.* iseatóipe
isotope effect éifeacht *f.* iseatóipe
isotropic antenna aeróg *f.* iseatrópach
isotropic gain neartú *m.* iseatrópach
iterative guidance treoraíocht *f.* atriallach

J

jamming plúchadh *m.*
Jeans mass mais *f.* Jeans
jerk sracadh *m.*
jet scaird *f.*
jet engine scairdinneall *m.*
jet propulsion scairdtiomáint *f.*
jet stream[1] (exhaust) scairdsceitheadh *m.*
jet stream[2] (current) scairdsruth *m.*
jettisonable inchaite thar bord
jettisonable belt crios *m.* inchaite thar bord
jettisonable fairing slíomú *m.* inchaite thar bord
jitter[1] (flicker or fluctuation in high frequency digital signal) pasghiodam *m.*; plúchadh *m.* raidió
jitter[2] (movement) giodam *m.*
Jovian planet pláinéad *m.* Ióbhach
Julian calendar féilire *m.* Iúil
Julian date dáta *m.* Iúil
Jupiter Iúpatar *m.*
Jupiter family (comet) clann *f.* Ióbhach (cóiméad)
Jupiter system córas *m.* Iúpatair

K

Kepler's law dlí *m.* Kepler
Keplerian orbit fithis *f.* Kepler
kick stage cic-chéim *f.*
kick-off mechanism meicníocht *f.* tosaigh
Kirkwood Gap bearna *f.* Kirkwood
kinetheodolite cinéiteodailít *f.*
kinetic energy storage unit aonad *m.* stórais fuinnimh chinéitigh
kinetic pressure brú *m.* cinéiteach
kinetic temperature teocht *f.* chinéiteach
kinetosis cinéatóis *f.*
knowledge database bunachar *m.* sonraí eolais
Kolmogorov's turbulence suaiteacht *f.* Kolmogorov

L

Lagrangian point pointe *m.* Lagrange
Lambert's law dlí *m.* Lambert
Lambert's problem fadhb *f.* Lambert
Lambertian surface dromchla *m.* Lambert
laminar flow sreabhadh *m.* lannach; sreabh *f.* lannach
laminar flow enclosure iniamh *m.* sreafa lannaigh
land station stáisiún *m.* talún
landing tuirlingt *f.*
landing gear gléasra *m.* tuirlingthe
lap belt crios *m.* coime
large space structure struchtúr *m.* mór spáis
large-scale structure (of the cosmos) struchtúr *m.* mórscála (an chosmais)
laser léasar *m.*
laser cooling fuarú *m.* léasair
laser-gyro gíreascóp *m.* léasair
last universal common ancestor [**LUCA**] an sinsearach *m.* coiteann uilíoch deiridh
latching valve comhla *f.* laiste
latent heat teas *m.* folaigh
latitude leithead *m.*
launch lainseáil *f.*
launch-and-entry-suit culaith *f.* lainseála agus iontrála
launch area láthair *f.* lainseála
launch base bunáit *f.* lainseála
launch complex coimpléasc *m.* lainseála
launch pad ceap *m.* lainseála
launch ramp rampa *m.* lainseála
launch ramp shelter scáthlán *m.* rampa lainseála
launch readiness review athbhreithniú *m.* ullmhachta lainseála
launch site ionad *m.* lainseála
launch table tábla *m.* lainseála
launch timetable amchlár *m.* lainseála

launch vehicle feithicil *f.* lainseála
launch window caoi *f.* lainseála; fuinneog *f.* lainseála
launcher lainseálaí *m.*
launching lainseáil *f.*
launching rail ráille *m.* lainseála
leading edge imeall *m.* tosaigh
leakage radiation radaíocht *f.* ligin
lean limit teorainn *f.* thanaí
leap day lá *m.* bisigh
leap second soicind *f.* bhisigh
left-hand circular polarization [**LCP**] polarú *m.* ciorclach tuathail
leg volume toirt *f.* coise
level sensor braiteoir *m.* leibhéil
levitation eadarbhuasú *m.*
libration guagadh *m.*
libration point pointe *m.* guagtha
LiDAR líodar *m.*
Liesegang phenomenon feiniméan *m.* Liesegang
life sciences eolaíochtaí *fpl.* beatha
life-support equipment trealamh *m.* tacaíochta beatha
life-support module modúl *m.* tacaíochta beatha
life-support system córas *m.* tacaíochta beatha
lifecycle cost [**LCC**] costas *m.* saolré
lifetime ré *f.*
lift ardú *m.*
lift-off éirí *m.*; ardú *m.* aníos
lift-off command ordú *m.* éirí
lift-off time am *m.* éirí
lifted re-entry athiontráil *f.* ardaithe
lifting body cabhail *f.* ardaithe
light cone cón *m.* solais
light-emitting diode [**LED**] dé-óid *f.* astaithe solais
light pen peann *m.* solais
light year [**Ly**] bliain *f.* solais
light-gas gun gunna *m.* gáis éadroim
limb imeall *m.*; géag *f.*
Lindblad resonance athshondas *m.* Lindblad
line of apsides líne *f.* na n-aipsisí

line of nodes líne *f.* nód
line spectrum speictream *m.* líne
linear absorptance ionsúiteas *m.* líneach
line-of-sight velocity treoluas *m.* líne amhairc
liner[1] **(lining)** líneáil *f.*
liner[2] **(airliner)** línéar *m.*
liquid bridge fracture experiment turgnamh *m.* scoilte droichid leachtaigh
liquid bridge Marangoni convection comhiompar *m.* Marangoni i ndroichead leachtach
liquid propellant tiomántán *m.* leachta
list (to) liostáil
listing liostú *m.*
lithergol liotargal *m.*
litho-Panspermia liota-panspeirmeachas *m.*
lithophile element dúil *f.* litifileach
little skirt sciorta *m.* beag
live beo
loading lódáil *f.*
lobar vibration creathadh *m.* lóbach
lobe lóba *m.*
Local Bubble boilgeog *f.* áitiúil
local cluster[1] **(local star cluster)** réaltbhraisle *f.* áitiúil; braisle *f.* áitiúil
local cluster[2] **(local galactic cluster)** réaltbhraisle *f.* réaltrach áitiúil; braisle *f.* réaltrach áitiúil
Local Interstellar Cloud scamall *m.* idir-réaltach áitiúil
local sidereal time réaltam *m.* áitiúil
local super-cluster ollbhraisle *f.* áitiúil
location suíomh *m.*
lock-on glasáil *f.* air
log-periodic antenna aeróg *f.* logartam-thréimhseach
long capillary method modh *m.* ribeadáin fhada
long-duration spaceflight spáseitilt *f.* fad-aga
longeron longarán *m.*
longitude fad *m.*
long-period comet cóiméad *m.* fadtréimhseach
look angle uillinn *f.* amhairc
loop swimming lúbshnámh *m.*

loran loran *m.*
loss-of-lock cailleadh *m.* glais
louver comhla *f.*
low lagrach *m.*; ísleán *m.*
low Earth orbit [LEO] fithis *f.* íseal Domhain
lower body negative pressure [LBNP] brú *m.* diúltach coirp íochtaraigh
lowest usable frequency [LUF] íosmhinicíocht *f.* inúsáidte
low orbit fithis *f.* íseal
low-pressure balloon balún *m.* lagbhrú
low-pressure system córas *m.* lagbhrú
low-surface brightness galaxy réaltra *m.* gile dromchla ísíl
low-temperature physics fisic *f.* ísealteochta
luminance lonras *m.*
luminosity lonrachas *m.*
luminosity distance fad *m.* lonrachais
luminosity function feidhm *f.* lonrachais
luminous exitance eisteachas *m.* lonrúil
luminous exposure nochtadh *m.* lonrúil
lunar crater cráitéar *m.* gealaí
lunar cycle ciogal *m.* na gealaí
lunar eclipse urú *m.* na gealaí
lunar meteorite dreigít *f.* ghealaí
lunar module modúl *m.* gealaí
lunar occultation folú *m.* na gealaí
lunar satellite satailít *f.* ghealaí
lunisolar calendar féilire *m.* ghealaí is ghréine
lunisolar precession luainíocht *f.* luanghrianda
Lyman-alpha cloud scamall *m.* Lyman-alpha
Lyman-alpha forest foraois *f.* Lyman-alpha
Lyman-alpha radiation radaíocht *f.* Lyman-alpha
Lyman break galaxy réaltra *m.* briseadh Lyman

M

Mach number uimhir *f.* Mach
MacIlwain parameter paraiméadar *m.* McIlwain
macroscopic quantum phenomena feiniméin *mpl.* mhacrascópacha chandaim
macrostep macraichéim *f.*
magnetic cleanliness glaineacht *f.* mhaighnéadach
magnetic flux tube feadán *m.* flosca mhaighnéadaigh
magnetic mirror effect éifeacht *f.* scátháin mhaighnéadaigh
magnetic mirror point pointe *m.* scátháin mhaighnéadaigh
magnetic reconnection athchónasc *m.* maighnéadach
magnetic shell sceall *m.* maighnéadach
magnetic storm stoirm *f.* mhaighnéadach
magnetic tail eireaball *m.* maighnéadach
magnetohydrodynamic propulsion tiomáint *f.* mhaighnéadaihidridinimiciúil
magnetohydrodynamics [**MHD**] maighnéadaihidridinimic *f.*
magnetopause maighnéadstad *m.*
magnetopause current sruth *m.* maighnéadstad
magnetoplasma maighnéada-plasma *m.*
magnetoplasmadynamic propulsion tiomáint *f.* mhaighnéada-plasmaidinimiciúil
magnetoplasmadynamics [**MPD**] maighnéada-plasmaidinimic *f.*
magnetosheath maighnéadchumhdach *m.*
magnetosphere maighnéadaisféar *m.*
magnetospheric bow-shock turraing *f.* thosaigh mhaighnéadaisféarach
magnetospheric cavity cuas *m.* maighnéadaisféarach
magnetospheric cavity-compression comhbhrú *m.* cuasáin mhaighnéadaisféaraigh
magnetospheric cleft gág *f.* mhaighnéadaisféarach
magnetospheric physics fisic *f.* mhaighnéadaisféarach
magnetospheric storm stoirm *f.* mhaighnéadaisféarach
magnetospheric substorm fostoirm *f.* mhaighnéadaisféarach

main body príomhchabhail *f.*
main sequence star réalta *f.* phríomhsheichimh
main-belt asteroid astaróideach *m.* príomhchreasa
maintainability inchotabhálacht *f.*
maintenance cotabháil *f.*
maintenance telemetry teiliméadracht *f.* chotabhála
major tone mórthon *m.*
manacle ring fáinne *m.* cuibhrigh; fáinne *m.* geimhle
man-machine integration system córas *m.* comhtháthaithe duine le meaisín
man-made noise torann *m.* saorga; torann *m.* de dhéantús an duine
mannable ináitrithe
manned le foireann; le fear
manned flight eitilt *f.* le foireann
manned maneuvring unit [**MMU**] aonad *m.* ionramhála le foireann
manned module modúl *m.* le foireann
manned spacecraft spásárthach *m.* le foireann
manned spaceflight spáseitilt *f.* le foireann
man-rated fear-rátaithe
many-body problem fadhb *f.* ilchorp
map mapa *m.*
map (**to**) mapáil
mapper mapaire *m.*
Marangoni convection comhiompar *m.* Marangoni
Marangoni number uimhir *f.* Marangoni
Mars Mars *m.*
Martian meteorite dreigít *f.* Mharsach
mascon mascan *m.*
maser méasar *m.*
mass-discharge coefficient comhéifeacht *f.* maisdíluchtaithe
mass function feidhm *f.* mhaise
massive compact halo object réad *m.* oll-dlúthluain
mass-luminosity relation coibhneas *m.* maise le lonracht
mass-transfer cooling fuarú *m.* maisaistrithe
matched nozzle soc *m.* comhoiriúnaithe
material experiment turgnamh *m.* ábhair
materials processing próiseáil *f.* ábhar
materials processing in space próiseáil *f.* ábhar sa spás

materials sciences eolaíochtaí *fpl.* na n-ábhar
Maxwell-Garnett approximation neastachán *m.* Maxwell-Garnett
mean albedo meán-ailbéideacht *f.*
mean anomaly meán-aimhrialtacht *f.*
mean downtime [MDT] meániamh-aga *m.*
mean longitude meánfhad *m.*
mean repair time [MRT] meánaga *m.* deisiúcháin
mean solar time meánghrian-am *m.*
mean time between failures [MTBF] meán-am *m.* idir teipeanna
mean time to failure [MTTF] meán-am *m.* go teip
mean time to restoration [MTTR] meán-am *m.* go hathchóiriú
mechanoperception meicneabhrath *m.*
mechanoreceptor meicneaghabhdóir *m.*
melt zone són *m.* leá; crios *m.* leá
memorandum of understanding [MOU] meabhrán *m.* tuisceana
Mercury Mearcair *m.*
meridian fadlíne *f.*
meridian plane plána *m.* fadlíne
meridian transit trasdul *m.* fadlíne
mesometeorology méisiméiteareolaíocht *f.*
mesopause méiseastad *m.*
mesosphere méisisféar *m.*
metabolism meitibileacht *f.*
metallicity miotalacht *f.*
metal matrix composites comhábhair *mpl.* mhaitríse miotail
metal-nitride-oxide semiconductor [MNOS] leathsheoltóir *m.* ocsaíd-nitríde miotail
metal-oxide semi-conductor [MOS] leathsheoltóir *m.* ocsaíde miotail
metastable liquid leacht *m.* meiteachobhsaí
metastable phase pas *m.* meiteachobhsaí
meteor dreige *f.*; meitéar *m.*; caor *f.* thine
meteor shower dreigechith *m.*
meteoric dreigeach; meitéarach; caorthineach
meteoric trail rian *m.* dreigeach
meteorite dreigít *f.*
meteorite impact tuinseamh *m.* dreigíte
meteoritic dreigíteach
meteoritic crater cráitéar *m.* dreigíteach

meteoritic tail eireaball *m.* dreigíteach
meteoroid dreigeoideach *m.*
meteoroid shield sciath *f.* dhreigeoidigh
meteorological optical range [MOR] raon *m.* optúil meitéareolaíoch
meteorological satellite satailít *f.* mheitéareolaíoch
meteorological visibility infheictheacht *f.* mheitéareolaíoch
meteorological-satellite service seirbhís *f.* satailíte meitéareolaíche
meteorology meitéareolaíocht *f.*
metric camera ceamara *m.* méadrach
microclimate miocraeráid *f.*
microdosimetry micreadháileogmhéadracht *f.*
microfossils micriontaisí *fpl.*
microgravity micrimtharraingt *f.*; micrea-dhomhantarraingt *f.*
microgravity engineering innealtóireacht *f.* mhicrimtharraingthe
microgravity processing próiseáil *f.* mhicrimtharraingthe
microgravity research taighde *m.* micrimtharraingthe
microgravity sciences eolaíochtaí *fpl.* micrimtharraingthe
micrometeorite micridhreigít *f.*
microprocessor micreaphróiseálaí *m.*
microstrip line líne *f.* mhicreastéille
microthruster micricheartaitheoir *m.*
microwave background cúlra *m.* micreathonnach
microwave band banda *m.* micreathonnach
microwave remote sensing cianbhraiteacht *f.* mhicreathonnach
microwave scatterometer scaipmhéadar *m.* micreathonnach
microwaves micreathonnta *fpl.*
mid-course correction manoeuvre ionramháil *f.* choigeartaithe meánchúrsa
military use of outer space úsáid *f.* mhíleata den spás amuigh
Milky Way Bealach *m.* na Bó Finne
Miller's experiment turgnamh *m.* Miller
million accounting units [MAU] milliún aonad *m.* cuntasaíochta
minor planet mionphláinéad *m.*
minor planet designation ainmniú *m.* mionphláinéid
minor tone mionton *m.*
Mira variable réalta *f.* athraitheach Mira
mirror furnace foirnéis *f.* scáthánach
mismatched nozzle soc *m.* mí-chomhoiriúnaithe

missile diúracán *m.*
mission misean *m.*
mission analysis anailís *f.* mhisin
mission control center lárionad *m.* rialaithe misin
mission telemetry teiliméadracht *f.* mhisin
mixing ratio cóimheas *m.* measctha
mixture composition comhshuíomh *m.* meascáin
mixture ratio cóimheas *m.* meascáin
mobile aeronautical station stáisiún *m.* aerloingseoireachta soghluaiste
mobile communications cumarsáid *f.* shoghluaiste
mobile land station stáisiún *m.* talún soghluaiste
mobile maritime station stáisiún *m.* mara soghluaiste
mobile satellite service seirbhís *f.* satailíte shoghluaiste
mobile station stáisiún *m.* soghluaiste
mock-up balbhshamhail *f.*; samhail *f.*
modal number líon *m.* módúil
mode of vibration mód *m.* creathaidh
mode shape cruth *m.* móid
model samhail *f.*; mionsamhail *f.*; múnla *m.*
modeling samhaltú *m.*; mionsamhlú *m.*
modified Julian date dáta *m.* Iúil mionathraithe
modular modúlach
modulation modhnú *m.*
modulation transfer function [**MTF**] feidhm *f.* aistriú modhnaithe
module modúl *m.*
Mohorovicic discontinuity neamhleanúnachas *m.* Mohorovicic
molecular aerodynamics aeraidinimic *f.* mhóilíneach
molecular cloud scamall *m.* móilíneach
molecular flow sreabhadh *m.* móilíneach; sreabh *f.* mhóilíneach
momentum wheel roth *m.* móimintim
monitor monatóir *m.*
monitor (to) monatóireacht a dhéanamh
monitoring monatóireacht *f.*
monolithic integrated circuit ciorcad *m.* comhtháite monailiotach
monopole antenna aeróg *f.* mhonapholach
monopropellant monathiomántán *m.*
monostatic radar radar *m.* monastatach
moon gealach *f.*

Moon An Ghealach *f.*
Moon's path (in celestial sphere) conair *f.* na Gealaí (i sféar neamhaí)
morphological instability éagobhsaíocht *f.* mhoirfeolaíoch
motor mótar *m.*
motor case cás *m.* mótair
Mullins-Sekerka theory teoiric *f.* Mullins-Sekerka
multi-beam antenna aeróg *f.* il-léasach
multi-channel ilchainéalach
multi-hop satellite link nasc *m.* satailíte ilphreabaí
multi-look ilradharc *m.*
multi-mission modular spacecraft [**MMS**] spásárthach *m.* modúlach ilmhisean
multiple access ilrochtain *f.*
multiple star ilréalta *f.*
multiplexer ilphléacsóir *m.*
multiplexing ilphléacsú *m.*
multiplier phototube fótafheadán *m.* iolrach
multipurpose ilchuspóireach
multi-satellite link nasc *m.* ilsatailíte
multi-spectral scanner [**MSS**] scanóir *m.* ilspeictreach
multi-spectral sensor braiteoir *m.* ilspeictreach
multi-stage launch vehicle feithicil *f.* lainseála ilchéime
multi-user facility rack raca *m.* saoráide ilúsáideoirí
muscle atrophy seargadh *m.* matáin
muscle loss caillteanas *m.* matáin
musculoskeletal system córas *m.* matánchnámharlaigh
mutagenesis mútaighineas *m.*

N

nadir nadair *f.*
narrow-line region réigiún *m.* caol-líne
natural convection comhiompar *m.* nádúrtha
natural environment timpeallacht *f.* nádúrtha
natural mode of vibration mód *m.* creathaidh nádúrtha
natural resources acmhainní *fpl.* nádúrtha
natural trajectory ruthag *m.* nádúrtha
nautical twilight breacsholas *m.* muirí
naval twilight breacsholas *m.* cabhlaigh
navigation loingseoireacht *f.*
navigation satellite satailít *m.* loingseoireachta
near-Earth object [**NEO**] réad *m.* neasdomhain
near-Earth asteroid [**NEA**] astaróideach *m.* neasdomhain
near-parabolic orbit fithis *f.* neaspharabóileach
nebular réaltnéalach
nephanalysis néalanailís *f.*
nephelogram neifiligram *m.*
nephelometer neifiliméadar *m.*
nephelometry neifiliméadracht *f.*
Neptune Neiptiún *m.*
Neptunian family clann *f.* Neiptiúnach
nervous system córas *m.* na néaróg
nested nozzle soc *m.* neadaithe
network líonra *m.*
network of stations líonra *m.* stáisiún
neuronal plasticity plaisteacht *f.* néarónach
neuro-sensory system córas *m.* néarchéadfach
neurovestibular system córas *m.* néarchuasánach
neutral body posture staidiúir *f.* neodrach coirp
neutral hydrogen gas gás *m.* neodrach hidrigine
neutral point pointe *m.* neodrach
neutral sheet brat *m.* neodrach

neutralizer neodraitheoir *m.*
neutron capture reaction imoibriú *m.* gabhála neodróin
neutron environment timpeallacht *f.* neodróin
neutron spectrum speictream *m.* neodróin
neutron star neodrónréalta *f.*
neutronography neodrónagrafaíocht *f.*
new calendar féilire *m.* nua
night-airglow aerloinnir *f.* oíche
night-glow breo *m.* oíche
nightside (of a planet) taobh *m.* oíche (de phláinéad)
noctilucent clouds scamaill *mpl.* oíchelonracha
nodal period tréimhse *f.* nódach
node nód *m.*
nodical elongation fadú *m.* nódach
noise torann *m.*; fothram *m.*
nominal ainmniúil
nominal orbital position suíomh *m.* fithiseach ainmniúil
non-competitive tender tairiscint *f.* neamhiomaíoch
non-compliance neamhchomhlíonadh *m.*
non-conformance neamhchomhfhreagairt *f.*
non-dilatable balloon balún *m.* neamhinleata
non-expanding balloon balún *m.* neamhfhairsingíoch
non-member state neamhbhallstát *m.*
non-optimum expansion (of a nozzle) fairsingiú *m.* neamhoptamach (soic)
non-regulated busbar [**NRB**] busbarra *m.* neamhrialaithe
non-static pressure brú *m.* neamhstatach
nosecone soc-chón *m.*
notching eangú *m.*
nova nóva *m.*
nozzle soc *m.*
nozzle efficiency éifeachtacht *f.* soic
nozzle insert ionsáiteán *m.* soic
nozzle ring fáinne *m.* soic
nozzle swiveling sclóineadh *m.* soic
nozzle throat scornach *f.* soic
nozzle-expansion area ratio cóimheas *m.* achar fairsingithe soic
nozzling blocáil *f.* aeraidinimiciúil

nuclear device gléas *m.* núicléach; feiste *f.* núicléach
nuclear emulsion eibleacht *f.* núicléach
nuclear energy fuinneamh *m.* núicléach
nuclear magnetic resonance [**NMR**] athshondas *m.* maighnéadach núicléach [AMN]
nuclear propulsion tiomáint *f.* núicléach
nuclear track detector brathadóir *m.* riain núicléach
nuclear-electric propulsion tiomáint *f.* núicléach-leictreach
nuclear-thermal propulsion tiomáint *f.* núicléach-teirmeach
nucleation núicléatú *m.*
nucleus núicléas *m.*
nudging broideadh *m.*
numeric uimhriúil
nutation nútú *m.*; nútaíocht *f.*
nutation damper maolaire *m.* nútaithe
nystagmus nisteagmas *m.*

O

object in space réad *m.* sa spás
obliquity of the ecliptic sceabhacht *f.* (an éicliptigh)
observation from space breathnóireacht *f.* ón spás; breathnú *m.* ón spás
observer breathnóir *m.*; breathnadóir *m.*
occultation folú *m.*
occupational medicine míochaine *f.* shaothair
oceanic tide taoide *f.* aigéanach
ocular counter-rolling frithrolladh *m.* súilí
oculogyral illusion seachmall *m.* ocalaigíreach
oculomotor response freagairt *f.* ocalamótair; freagairt *f.* shúil luadrach
off-line as líne
off-gassing seachghasáil *f.*
offset feed fotha *m.* seach-chuir
OH maser méasar *m.* OH
omnidirectional antenna aeróg *f.* uilethreoch
omnidirectional range raon *m.* uilethreoch
onboard computer [**OBC**] ríomhaire *m.* ar bord
onboard equipment trealamh *m.* ar bord
onion structure struchtúr *m.* oinniúin
on-line ar líne
on-off control rialú *m.* ann as
on-orbit (satellite) ar fhithis (satailít)
on-orbit assembly cóimeáil *f.* ar fhithis
on-orbit repair deisiúchán *m.* ar fhithis
on-station (satellite) ar shuíomh; ar stáisiún (satailít)
Oort limit teorainn *f.* Oort
Oort's cloud scamall *m.* Oort
open balloon balún *m.* oscailte
open cluster réaltbhraisle *f.* oscailte; braisle *f.* oscailte
open competitive tender tairiscint *f.* iomaíoch oscailte
open universe cruinne *f.* oscailte
open-neck balloon balún *m.* scornaí oscailte

operate (to) oibriú
operation oibriú *m.*; oibriúchán *m.*; oibríocht *f.*
operational maximum usable frequency uasmhinicíocht *f.* inúsáidte oibríoch
operational phase pas *m.* oibríoch
operational space medicine míochaine *f.* oibríochtúil spástaistil
operations phase pas *m.* na n-oibríochtaí
opposition urchomhaireacht *f.*
optical detector brathadóir *m.* optúil
optimization optamú *m.*
optimum expansion (of a nozzle) fairsingiú *m.* optamach (soic)
optokinetic after-nystagmus iarnisteagmas *m.* optaicinéiteach
optokinetic nystagmus nisteagmas *m.* optaicinéiteach
optokinetic system córas *m.* optaicinéiteach
orbit fithis *f.*
orbit (to) fithisiú
orbit decay meath *m.* fithise
orbit determination cinneadh *m.* fithise
orbit inclination claonadh *m.* fithise
orbit injection instealladh *m.* fithise
orbit insertion ionsá *m.* fithise
orbit lowering ísliú *m.* fithise
orbital base bunáit *f.* fhithiseach
orbital debris smionagar *m.* fithiseach
orbital elements eilimintí *fpl.* fithise
orbital flight eitilt *f.* fhithiseach
orbital lifetime ré *f.* fhithiseach
orbital manoeuvre ionramháil *f.* fhithiseach
orbital period tréimhse *f.* fhithiseach
orbital station stáisiún *m.* fithiseach
orbital transfer vehicle [OTV] feithicil *f.* aistrithe fhithiseach
orbital weapon arm *m.* fithiseach
orbiter fithiseoir *m.*
orbiting fithisiú
orbiting object réad *m.* fithisitheach
orbitography fithiseagrafaíocht *f.*
ordinary chondrite gnáthchoindrít *f.*
organigramme plean *m.* eagraíochta

organization chart cairt *f.* eagraíochta
origin of life bunús *m.* na beatha
orthostatic hypotension fotheannas *m.* ortastatach
orthostatic-hypotensive intolerance éadulaingt *f.* ortastatach fhotheannasach
orthostatic intolerance éadulaingt *f.* ortastatach
orthostatic tolerance fulaingt *f.* ortastatach
orthostatism ortastatachas *m.*
oscillating environment timpeallacht *f.* ascalach
oscillatory Marangoni convection comhiompar *m.* Marangoni ascalach
oscillopsia ascalaipsia *f.*
osculating orbital elements eilimintí *fpl.* fithise sárthadhlacha
osteoblast oistéablast *m.*
osteogenesis oistéigineas *m.*
osteopenia oistéipéinia *f.*
osteoporosis oistéapóróis *f.*
otocyst otaicist *f.*
otolith otailit *f.*
otolith degeneration meathlú *m.* otailite
otolith ocular response freagairt *f.* shúilí otailite
otolith organ ball *m.* otailite; orgán *m.* otailite
otolith tilt-translation reinterpretation [**OTTR**] athléirmhíniú *m.* claonaistriúcháin otailite
outage idirscor *m.*; briseadh *m.*
outer planet pláinéad *m.* lasmuigh
outer solar system grianchóras *m.* lasmuigh
outer space an spás *m.* amuigh
out-gassing sceitheadh *m.* gáis; asgasáil *f.*
over-expanded ró-fhairsingthe
over-expansion ró-fhairsingiú *m.*
over-pressure ró-bhrú *m.*
over-the-horizon propagation forleathadh *m.* trasléaslíne
oxidant ocsaídeán *m.*
oxidizer ocsaídeoir *m.*
oxygen mask masc *m.* ocsaigine
ozonosphere ozónaisféar *m.*

P

packaging pacáistiú *m.*; pacáistíocht *f.*
pallet pailléad *m.*
panspermia panspeirmeachas *m.*
parabolic flight eitilt *f.* pharabóileach
parabolic orbit fithis *f.* pharabóileach
parabolic reflector frithchaiteoir *m.* parabóileach
paraboloid antenna aeróg *f.* pharabólóideach
parallax saobhdhiallas *m.*
parawing eite *f.* pharaisiúit
parking orbit fithis *f.* pháirceála
parsec [pc] parsoic *f.*
particle dispersion idirleathadh *m.* cáithnín
particle drift síobadh *m.* cáithnín
passenger paisinéir *m.*
passive éighníomhach
passive satellite satailít *f.* éighníomhach
passive sensor braiteoir *m.* éighníomhach
patched conics technique teicníocht *f.* cónghearrthacha paisteálta
patent right ceart *m.* paitinne
pattern detection brath *m.* patrúin
payload pálasta *m.*
payload accommodation oiriúntas *m.* pálasta
payload specialist speisialtóir *m.* pálasta
peculiar minor planet mionphláinéad *m.* corr
peculiar motion gluaisne *f.* dhílis
peg formation foirmiú *m.* peige (meá-mhorfaigineas)
pencil-beam antenna aeróg *f.* léas pionsail
penguin suit culaith *f.* phiongaine
penumbra leathscáil *f.*; páirtscáil *f.*
performance efficiency éifeachtacht *f.* feidhmiúcháin
periapsis peiriaipsis *f.*; neasphointe *m.* fithise
periastron peiriastrón *m.*

perigee peirigí *m.*
perigee boost motor mótar *m.* treisithe peirigí
perigee rocket roicéad *m.* peirigí
perigee stage céim *f.* pheirigí
perihelion peirihéilean *m.*
period of revolution tréimhse *f.* imrothlaithe
permafrost síorshioc *m.*
permanent boost treisiú *m.* buan
personal pearsanta
perturbation corraíl *f.*
perturbation equation cothromóid *f.* chorraíola
perturbed orbit fithis *f.* chorraíola
perturbing acceleration luasghéarú *m.* corraitheach
phantom torso colainn *f.* taibhse
phase A pas A
phase angle pasuillinn *f.*; uillinn *f.* phas
phase B pas B
phase C pas C
phase D pas D
phase integral pas-suimeálaí *m.*
phase interference pastrasnaíocht *f.*
phase jump pasléim *f.*
phase modulation [**PM**] pasmhodhnúchán *m.* [PM]
phase of a space project pas *m.* de thionscadal spáis
phase selection pasroghnú *m.*
phase separation pasdeighilt *f.*
phase skip pas-scip *f.*
phase transition pas-trasdul *m.*
phased-array (**antenna**) paseagar *m.* aeróg
phase-locked loop [**PLL**] lúb *f.* phasghlasáilte
phase-shift keying [**PSK**] eochrú *m.* pasaistrithe
phoresis separation deighilt *f.* fóiréise
photo-cell fótaichill *f.*
photo-detection fótóbhrath *m.*
photo-dissociation region réigiún *m.* fótái-dhíthiomsúcháin
photo-electric cell cill *f.* fhótaileictreach
photogrammetric camera ceamara *m.* fótagraiméadrachta
photometry fótaimhéadracht *f.*

photon propulsion tiomáint *f.* fótón
photosphere fótaisféar *m.*
photo-voltaic cell cill *f.* fhótavoltach
phytotron fíteatrón *m.*
pick-a-back satellite satailít *f.* heigh-deá
piggyback satellite satailít *f.* muin muice
piston effect éifeacht *f.* loine
pitch céim *f.*
pitch angle céim *f.* uillinn; uillinn *f.* chéime
pixel picteilín *m.*
place in orbit (to) cur i bhfithis
Planck length fad *m.* Planck
Planck time am *m.* Planck
plane albedo ailbéideacht *f.* phlána
planet pláinéad *m.*
planetary pláinéadach
planetary biology bitheolaíocht *f.* phláinéadach
planetary boundary layer sraith *f.* teorann plainéadaí
planetary defense cosaint *f.* phláinéadach
planetary disk diosca *m.* pláinéadach
planetary evolution éabhlóid *f.* phláinéadach
planetary friction layer sraith *f.* frithchuimilte phláinéadach
planetary geology geolaíocht *f.* phláinéadach
planetary lander landálaí *m.* pláinéadach
planetary mission misean *m.* pláinéadach
planetary nebula réaltnéal *m.* pláinéadach
planetary orbiter fithiseoir *m.* pláinéadach
planetary precession luainíocht *f.* phláinéadach
planetary probe taiscéalaí *m.* pláinéadach
planetary protection cosaint *f.* phláinéadach
planetary rover carr *m.* pláinéadach
planetary simulation chamber seomra *m.* ionsamhlaithe pláinéadach; cuasán *m.* ionsamhlaithe pláinéadach
planetesimal mionphláinéadach
planetoid pláinéadóideach *m.*
planetology pláinéadeolaíocht *f.*
plasma plasma *m.*
plasma frequency minicíocht *f.* phlasma

plasma propulsion tiomáint *f.* phlasma
plasma sheet brat *m.* plasma
plasmapause plasmastad *m.*
plasmasphere plasmaisféar *m.*
plastic track detectors brathadóirí *mpl.* rianáin phlaistigh
plate tectonics teicteonaic *f.* phlátaí
platform ardán *m.*
plotter breacaire *m.*
plug nozzle soc *m.* stopalláin
plume cleitín *m.*
Pluto Plútón *m.*
plutoid plútóideach *m.*
pogo effect éifeacht *f.* pógó
pointing accuracy beachtas *m.* pointeála
polar aurora gealáin *mpl.* pholacha
polar cap caidhp *f.* pholach
polar cusp cuspa *m.* polach
polar motion gluaisne *f.* pholach
polar mount taca *m.* polach
polar movement gluaiseacht *f.* pholach
polar orbit fithis *f.* pholach
polar region réigiún *m.* polach
polar transport iompar *m.* polach
Polaris Polaris *m.*; gréasán *m.* Polaris; an réalta *f.* thuaidh
polarization polárú *m.*
pollution truailliú *m.*
polycyclic aromatic hydrocarbons [PAH] hidreacarbóin *mpl.* aramatacha iltimthrialacha
population I star réaltchineál *m.* I
population II star réaltchineál *m.* II
population III star réaltchineál *m.* III
port-to-throat area ratio cóimheas *m.* achair port go scornach
position determination cinneadh *m.* suímh
position finding aimsiú *m.* suímh
position fixing fosú *m.* suímh
positioning suí *m.*
positioning phase pas *m.* suite
post-flight analysis anailís *f.* iareitilte

post-flight orthostatic intolerance éadulaingt *f.* ortastatach iareitilte
postural muscle matán *m.* staidiúireach
potting potáil *f.*
power conditioner feabhsaitheoir *m.* cumhachta
power flux density [**PFD**] floscdhlús *m.* cumhachta
power module modúl *m.* cumhachta
power spectral density [**PSD**] dlús *m.* speictreach cumhachta
power system córas *m.* cumhachta
powered balloon balún *m.* faoi thiomáint
powered flight eitilt *f.* faoi thiomáint
Poynting-Robertson effect éifeacht *f.* Poynting-Robertson
pre-biotic chemistry ceimic *f.* réamhbhitheach
precession luainíocht *f.*
precipitation frasaíocht *f.*; deascadh *m.*
pre-emphasis réamhbhéimniú *m.*
pre-flight adaptation training [**PAT**] traenáil *f.* oirúnaithe réamheitilte
pre-flight centrifuge training traenáil *f.* lártheifneora réamheitilte
preliminary analysis phase pas *m.* réamhanailíse
preliminary design review [**PDR**] athbhreithniú *m.* réamhdhearaidh
pre-processing réamhphróiseáil *f.*
pressing preasáil *f.*; brú *m.*
pressure breathing system córas *m.* análaithe faoi bhrú
pressure ratio cóimheas *m.* brú
pressurization brú-chóiriú *m.*
pressurization testing tástáil *f.* bhrú-chóirithe
pressurized suit culaith *f.* bhrú-chóirithe
pressurizing gas gás *m.* brú-chóiritheach
preventive maintenance cotabháil *f.* choisctheach
primacord primacord *m.*; corda *m.* tarchuir
primary battery bataire *m.* príomhúil
primary cell cillín *m.* príomhúil
primary radar radar *m.* príomhúil
prime contractor conraitheoir *m.* príomha
priming prímeáil *f.*
primordial atmosphere atmaisféar *m.* príomhordúil
primordial solar nebula réaltnéal *m.* gréine príomhordúil
print out (to) priontáil amach
printout prionta *m.* amach

procurement proposal togra *m.* soláthair
product táirge *m.*; toradh *m.*
product assurance dearbhú *m.* táirge
production phase pas *m.* táirgthe
profile próifíl *f.*
project manager bainisteoir *m.* tionscadail
project review athbhreithniú *m.* tionscadail
project team foireann *f.* tionscadail
prominence starraiceacht *f.*
proof pressure testing tástáil *f.* brú promhaidh
propellant tiomántán *m.*
propellant grain gráinne *m.* tiomántáin
propellant mass mais *f.* tiomántáin
propellant with metallic additive tiomantán *m.* le breiseán miotalach
proper motion dualghluaisne *f.*
property rights cearta *mpl.* maoine
proposal togra *m.*
proposal evaluation board bord *m.* measúnaithe togra
proposal opening board bord *m.* oscailte togra
proprioception féinghabhdóireacht *f.*
proprioceptive stimulation spreagadh *m.* féinghabhdóireachta
proprioceptor féinghabhdóir *m.*
propulsion tiomáint *f.*
propulsion bay bá *f.* thiomána
propulsion system córas *m.* tiomána
propulsion system manufacturer déantúsóir *m.* córais tiomána
propulsive flow sreabhadh *m.* tiomána; sreabh *f.* tiomána
protection ratio cóimheas *m.* cosanta
protective clothing éadach *m.* cosanta
protocol prótacal *m.*
proto-flight model samhail *f.* prótaeitilte
proto-galaxy prótairéaltra *m.*
proto-planet prótapláinéad *m.*
proto-planetary disk diosca *m.* prótapláinéadach
proto-star prótairéalta *f.*
prototype fréamhshamhail *f.*
psychological stress brú *m.* síceolaíoch
psychology síceolaíocht *f.*

psychomotor ability cumas *m.* sícealuaile
psychomotor performance éifeachtacht *f.* sícealuaile
Ptolemaic theory teoiric *f.* Tholaméis
publication of scientific data foilsiú *m.* sonrái eolaíochta
puffy face aghaidh *f.* shéidte
puffy face syndrome siondróm *m.* aghaidh shéidte
pulsar pulsár *m.*
pulsar dispersion spré *m.* pulsáir
pulsating variable réalta *f.* athraitheach bhíogach
pulse jet engine scairdinneall *m.* bíog
pulse-amplitude modulation [**PAM**] modhnú *m.* aimplitiúide bíog
pulse-code modulation [**PCM**] modhnú *m.* bíogchóid
pulse-duration modulation [**PDM**] modhnú *m.* bíogfhaid
pulse-modulation modhnú *m.* bíog
pulse-position modulation [**PPM**] modhnú *m.* bíogionaid
pulse-width modulation [**PWM**] modhnú *m.* bíogleithid
pultrusion tarraing-easbhrú *m.*
pupil mac *m.* imrisc
purpose aidhm *f.*; cuspóir *m.*
push-broom sensor braiteoir *m.* brúscuaibe
pyranometer pireanaiméadar *m.*
pyromechanism pirimeicníocht *f.*
pyrotechnic piriteicniúil
pyrotechnic chain slabhra *m.* piriteicniúil
pyrotechnic composition comhshuíomh *m.* piriteicniúil
pyrotechnic train sraith *f.* phiriteicniúil
pyrotechnics piriteicnic *f.*

Q

quadrature cearnaíocht *f.*; cearnú *m.*
qualification cáilíocht *f.*
qualification model [QM] samhail *f.* cháilíochta
qualification test tástáil *f.* cháilíochta
quality cáilíocht *f.*
quality assurance dearbhú *m.* cáilíochta
quality management bainistíocht *f.* cáilíochta
quality manual lámhleabhar *m.* cáilíochta
quality plan (of a product) plean *m.* cáilíochta (táirge)
quantization candamú *m.*; candamúchán *m.*
quantum nucleation núicléatú *m.* candamach
quasar cuasár *m.*
quasar absorption system córas *m.* ionsúcháin cuasáir
quasi-parabolic orbit fithis *f.* chuasapharabólach
quenching fuarú *m.*
quick-look mearamharc *m.*
quiet sun grian *f.* shuaimhneach
quotient líon *m.*

R

rack raca *m.*

radar radar *m.*

radar astronomy réalteolaíocht *f.* radair

radar scatterometer scaipmhéadar *m.* radair

radar altimeter airdemhéadar *m.* radair

radial velocity treoluas *m.* gathach

radiance radantas *m.*; loinnir *f.*

radiant (point) radaí *m.* (pointe)

radiant exposure nochtadh *m.* radanta

radiation radaíocht *f.*

radiation balance cothromaíocht *f.* radaíochta

radiation belt crios *m.* radaíochta

radiation biology bitheolaíocht *f.* radaíochta; radaibhitheolaíocht *f.*

radiation budget buiséad *m.* radaíochta

radiation diagram léaráid *f.* radaíochta

radiation hazard guais *f.* radaíochta

radiation noise torann *m.* radaíochta

radiation pattern patrún *m.* radaíochta

radiation protection cosaint *f.* radaíochta

radiation risk riosca *m.* radaíochta

radiation susceptibility so-ghabhálacht *f.* radaíochta

radiative cooling fuarú *m.* radantach

radioactive propellant tiomántán *m.* radaighníomhach

radio altimeter airdemhéadar *m.* raidió

radio astronomy radairéalteolaíocht *f.*

radio beacon rabhchán *m.* raidió

radio burst rois *f.* raidió

radio-communications cumarsáid *f.* raidió

radio-communications service seirbhís *f.* cumarsáide raidió

radio-determination radaichinneadh *m.*

radio-determination by satellite radaichinneadh *m.* le satailít

radio-frequency channel cainéal *m.* radaimhinicíochta

radio-frequency visibility infheictheacht *f.* radaimhinicíochta
radio-frequency [**RF**] radaimhinicíocht *f.*; minicíocht *f.* raidió
radio-galaxy radairéaltra *m.*
radio horizon léaslíne *f.* raidió
radio interference trasnaíocht *f.* raidió
radio interferometer rada-thrasnamhéadar *m.*
radio jet scaird *f.* raidió
radio link nasc *m.* raidió
radio lobe lóba *m.* raidió
radiometer radaimhéadar *m.*
radiometric albedo ailbéideacht *f.* radaiméadrach
radiometric resolution taifeach *m.* radaiméadrach
radio-navigation radaloingseoireacht *f.*
radio-paging glaoireacht *f.* raidió
radio-position-finding aimsiú *m.* suímh raidió
radio-protector radachosantóir *m.*
radio-sonde raidió *m.* sondála
radio-source radafhoinse *f.*
radio-telescope radaiteileascóp *m.*
radiowave radathonn *f.*
radome cón *m.* radair
ramjet rop-scairdeitleán *m.*
ramjet engine rop-scairdinneall *m.*
ramrocket roproicéad *m.*
random access rochtain *f.* randamach; randamrochtain *f.*
random noise torann *m.* randamach
random positioning machine meaisín *m.* suite randamaigh
random-access memory [**RAM**] cuimhne *f.* randamrochtana
range[1] (**of activities**) raon *m.* gníomhachtaí
range[2] (**distance**) raon *m.*
range[3] (**testing**) láthair *f.* thástála
range-finder raonaimsitheoir *m.*
range-finding raonaimsiú *m.*
range-rate raonráta *m.*; treoluas *m.* gathach
range-rate measurement tomhas *m.* raonráta
ranging raonáil *f.*
rapid solidification mearsholadú *m.*
raster rastar *m.*; gréasán *m.*

rat hind-limb suspension crochadh *m.* ghéag deiridh luchóige móire
rate ráta *m.*
rate gyro rátaghíorascóp *m.*
rate-integrating gyro rátaghíorascóp *m.* suimeála
rated capacity toilleadh *m.* rátaithe
ratio cóimheas *m.*
Rayleigh-Bénard convection comhiompar *m.* Rayleigh-Bénard
Rayleigh-Gans approximation neastachán *m.* Rayleigh-Gans
Rayleigh-Taylor instability éagobhsaíocht *f.* Rayleigh-Taylor
re-adaptation athoiriúnú *m.*
re-entry athiontráil *f.*
re-entry angle uillinn *f.* athiontrála
re-entry capsule capsúl *m.* athiontrála
re-entry corridor conair *f.* athiontrála
re-entry vehicle feithicil *f.* athiontrála
reaction engine inneall *m.* imoibriúcháin
reaction propulsion tiomáint *f.* imoibriúcháin
reaction sphere sféar *m.* imoibriúcháin
reaction wheel roth *m.* imoibriúcháin
reaction-diffusion system córas *m.* imoibriúcháin agus idirleata
reactor core croí *m.* imoibreora
read-only memory [ROM] cuimhne *f.* inléite amháin
real time fíor-am *m.*
rear skirt sciorta *m.* cúil; sciorta *m.* deiridh
reattachment athcheangal *m.*
re-breather athanáileoir *m.*
recalescence athchalras *m.*
receiver glacadóir *m.*
recognition aithint *f.*
recombination line líne *f.* athchuingrithe
recondition athriochtú *m.*
reconfigurable in-athchumraithe
reconnaissance satellite satailít *f.* taiscéalaíochta
recoverable capsule capsúl *m.* in-athfhála
recovery athfháil *f.*
recovery capsule capsúl *m.* athfhála
recovery capsule experiment turgnamh *m.* capsúil athfhála
recovery factor fachtóir *m.* athfhála

red dwarf star abhacréalta *f.* dhearg
red giant star fathachréalta *f.* dhearg
red shift deargaistriú *m.*
redundancy iomarcaíocht *f.*
reference atmosphere atmaisféar *m.* tagartha
reflectance frithchaiteas *m.*
reflecting satellite satailít *f.* fhrithchaiteach
reflection nebula réaltnéal *m.* frithchaithimh
reflectivity frithchaiteacht *f.*
refraction athraonadh *m.*
refractory element eilimint *f.* shártheasfhulangach
refrigerant cuaisneán *m.*
refrigeration cuisniú *m.*; cuisniúchán *m.*
refueling athbhreoslú *m.*
regenerative cooling fuarú *m.* athghiniúnach
regenerative repeater athsheoltóir *m.* athghiniúnach
registered channels cainéil *mpl.* chláraithe
registration of space objects clárú *m.* réad spás
regulated busbar [RB] busbarra *m.* rialaithe
regulation rialúchán *m.*; rialachán *m.*; rialú *m.*
relative gain neartú *m.* coibhneasta; neartúchán *m.* coibhneasta
relative wind gaoth *f.* choibhneasta
release jaws gialla *mpl.* bíse
reliability iontaofacht *f.*
relict radiation fuíollradaíocht *f.*
remodeling athshamhaltú *m.*
remote control cianrialú *m.*
remote maintenance cianchotabháil *f.*
remote manipulator arm cianghéag *f.* ionramhálaí
remote sensing cianbhraiteacht *f.*
remote sensing satellite satailít *f.* chianbhraiteachta
remote sensor cianbhraiteoir *m.*
remote sounding ciansondáil *f.*
rendezvous coinne *f.*
repair deisiú *m.*; deisiúchán *m.*
repeater athsheoltóir *m.*
report tuairisc *f.*; tuarascáil *f.*
rescue tarrtháil *f.*

research and development [**R&D**] taighde *m.* agus forbairt
residual gravity imtharraingt *f.* iarmharach
residual propellant tiomántán *m.* iarmharach
resolution taifeach *m.*
resources management bainistíocht *f.* acmhainní
response freagairt *f.*; freagra *m.*
restricted three-body problem fadhb *f.* trí chorp theoranta
retrograde orbit fithis *f.* chasiompaithe
retro-rocket roicéad *m.* moillithe
return factor fachtóir *m.* aisfhillte
return vehicle feithicil *f.* fhillte
reusable in-athúsáidte
review debriefing meeting cruinniú *m.* dífhaisnéise athbhreithnithe
review team foireann *f.* athbhreithnithe
revolution iomrothlú *m.*; casadh *m.*
rib easna *f.*
Riemannian geometry céimseata *f.* Riemann
right ascension dronairde *f.*
right-hand circular polarization [**RCP**] polárú *m.* ciorclach deisil
righting behavior iompar *m.* ceartaithe
ring current sruth *m.* fáinneach
ring-laser gyro [**RLG**] gíorascóp *m.* léasar fáinneach
riometer riaiméadar *m.*
ripple tonnán *m.*
risk riosca *m.*; priacal *m.*
RNA world domhan *m.* RNA
Robertson-Walker metric méadrach *m.* Robertson-Walker
robotic arm géag *f.* róbatach
robotic mission misean *m.* róbatach
robotics róbataic
Roche radius raidias *m.* Roche; ga *m.* Roche
rocket roicéad *m.*
rocket (combustion) motor mótar *m.* (dó) roicéid
rocket engine inneall *m.* roicéid
rocket experiment turgnamh *m.* roicéid
rocket propulsion tiomáint *f.* roicéid
rocketry roicéadacht *f.*
roll rolla *m.*; guairneáil *f.* (satailíte)

roll-out rolladh *m.* amach
rotating habitat gnáthóg *f.* rothlach
rotating-wall vessel soitheach *m.* balla rothlaigh
rotation rothlú *m.*; rothlúchán *m.*
rotor rótar *m.*
roundtrip time aga *m.* dé-aistir; am *m.* dé-aistir
RR Lyrae variable réalta *f.* athraitheach RR Lyrae

S

sacculus sacalas *m.*
Sachs-Wolfe effect éifeacht *f.* Sachs-Wolfe
safety sábháilteacht *f.*
safety belt crios *m.* sábhála
safety program clár *m.* sábháilteachta
sample sampla *m.*
sampling sampláil *f.*
sampling interval eatramh *m.* samplála
satellite satailít *f.*
satellite broadcasting craoladh *m.* satailíte
satellite constellation buíon *f.* satailítí
satellite link nasc *m.* satailíte
satellite network líonra *m.* satailíte
satellite system córas *m.* satailíte
satellite-based satailítbhunaithe
Saturn Satarn *m.*
scan swath sraith *f.* scanta
scanner scanóir *m.*
scanning scanadh *m.*
scanning multichannel microwave radiometer [**SMMR**] radaimhéadar *m.* micreathonnach ilchainéal scanacháin
scattered wave tonn *f.* scaipthe
scattering scaipeadh *m.*
scatterometer scaipmhéadar *m.*
scene radharc *m.*
Scheil's equation cothromóid *f.* Scheil
Schuler pendulum luascadán *m.* Schuler
Schwarzschild radius raidias *m.* Schwarzschild; ga *m.* Schwarzschild
scrambler scrobhóir *m.*
scramjet scramscairdinneall *m.*; ropscairdeitleán *m.* dóchain fhorshonaigh
screened room seomra *m.* sciata
screening (test) tástáil *f.* sciata

scrubber sciúrthóir *m.*
sea-ice reactor imoibreoir *m.* muiroighir
sea-truth fírinne *f.* na mara
seal séala *m.*
sealing séalú *m.*
search for extraterrestrial intelligence [**SETI**] cuardach *m.* intleachta eachtardhomhanda
seat belt crios *m.* suíocháin; crios *m.* sábhála
second cosmic velocity an dara treoluas *m.* cosmach
secondary battery taisc-cheallra *m.*; taiscbhataire *m.*
secondary radar radar *m.* tánaisteach
secular perturbation corraíl *f.* chiantréimhseach
secular resonance athshondas *m.* ciantréimhseach
segmented rocket engine inneall *m.* roicéid deighilte
segregation leithscaradh *m.*
selective extinction díobhadh *m.* roghnaíoch
semi-ballistic re-entry athiontráil *f.* leathbhalaistíoch
semicircular canals canálacha *fpl.* leathchiorclacha
semiconductor leathsheoltóir *m.*
sensitivity íogaireacht *f.*
sensor braiteoir *m.*
sensori-motor system córas *m.* luaile céadfaí
sensory conflict coinbhleacht *f.* chéadfach; mearbhall *m.* mothúchán
separation deighilt *f.*
separation bay bá *f.* dheighilte
separation experiment turgnamh *m.* deighilte
sequence seicheamh *m.*
sequencer seicheamhóir *m.*
sequential access rochtain *f.* sheicheamhach
sequential test tástáil *f.* sheicheamhach
serial access rochtain *f.* shrathach
service seirbhís *f.*
service arc stua *m.* seirbhíse
service area limistéar *m.* seirbhíse
service module modúl *m.* seirbhíse
service tower túr *m.* seirbhíse
setting socrú *m.*
settlement on celestial bodies lonnaíocht *f.* ar reanna neimhe

Seyfert galaxy réaltra *m.* Seyfert
shannon seanán *m.*
shaped-beam antenna aeróg *f.* léas cruthach
shear-cell method modh *m.* fiarchille
shear-spinning fiarchasadh *m.*; fiarghuairneáil *f.*
sheathed detonating cord corda *m.* maidhmitheach cumhdaithe
shepherd satellite satailít *f.* aoire
shielded room seomra *m.* sciata
shielding sciathadh *m.*
shock-tube feadán *m.* turrainge
short-arm centrifuge lártheifneoir *m.* gearrghéige
short-duration microgravity experiment turgnamh *m.* micrimtharraingthe gearr-aga
short-period comet cóiméad *m.* gearrthréimhseach
shoulder harness úim *f.* ghualainne
shroud cumhdach *m.*
sideband taobh-bhanda *m.*
side-looking airborne radar [SLAR] radar *m.* aerbheirthe taobhamhairc
side-looking radar [SLR] radar *m.* taobhamhairc
sidereal day réaltlá *m.*
sidereal period of revolution réalt-tréimhse *f.* iomlaithe
sidereal time réaltam *m.*
sidereal year réaltbhliain *f.*
siderite sidirít *f.*; dreigít *f.* iarainn
siderolite sidirilít *f.*
siderophile element eilimint *f.* shidirifileach
signal comhartha *m.*
signal processor próiseálaí *m.* comharthaí
Silk damping maolú *m.* Silk
simulated microgravity micrimtharraingt *f.* ionsamhalta
simulation insamhladh *m.*
simulator ionsamhlóir *m.*
single-base propellant tiomántán *m.* aonbhun
single channel per carrier [SCPC] aonchainéal *m.* an t-iompróir
single-point failure [SPF] teip *f.* aonphointe
single sideband [SSB] taobh-bhanda *m.* singil
SiO maser méasar *m.* SiO
site ionad *m.*; láthair *f.*; suíomh *m.*

sizing socrú *m.* méide
skeletal calcium loss caillteanas *m.* cailciam cnámharlaigh
skin craiceann *m.*
skirt sciorta *m.*
sky background cúlra *m.* spéire
sled carr *m.* sleamhnáin
sleep deprivation díth *f.* codlata; easpa *f.* codlata
sloshing luascadh *m.*
slot antenna aeróg *f.* shliotáin
slow-twitch muscles matáin *mpl.* mhallbhíogacha
slush propellant tiomantán *m.* greallaí
small solar system body rinn *m.* grianchórais bhig
snatch-off connector nascaire *m.* snaptha as
soft landing tuirlingt *f.* bhog
soft mock-up balbhshamhail *m.* bhog
software bogearraí *mpl.*
software development cycle cioclad *m.* forbartha bogearraí
software engineering innealtóireacht *f.* bogearraí
software lifecycle saolré *f.* bogearraí
software package pacáiste *m.* bogearraí
solar activity grianghníomhaíocht *f.*
solar activity index innéacs *m.* grianghníomhaíochta
solar array grianeagar *m.*
solar battery grianchealllra *m.*
solar burst rois *f.* ghréine
solar calendar féilire *m.* na gréine
solar cell grianchill *f.*
solar cell array eagar *m.* grianchille
solar composition comhshuíomh *m.* na gréine
solar constant griantairiseach *m.*
solar corona grianchoróin *f.*
solar cosmic rays [SCR] grianghathanna *mpl.* cosmacha
solar cycle grianchioclad *m.*
solar eclipse urú *m.* na gréine
solar flare grianbhladhm *f.*
solar gamma-ray gáma-gha *m.* gréine
solar nebula réaltnéal *m.* na gréine
solar neutrino grian-neoidríonó *m.*

solar panel grianphainéal *m.*; painéal *m.* gréine
solar particle event [SPE] teagmhas *m.* cáithnín gréine
solar power station stáisiún *m.* grianchumhachta
solar propulsion tiomáint *f.* ghréine
solar sail grianseol *m.*
solar sail propulsion tiomáint *f.* ghrianseoil
solar spectrum grianspeictream *m.*
solar system grianchóras *m.*
solar UV radiation radaíocht *f.* gréine UV
solar wind grianghaoth *f.*
solar-electric propulsion tiomáint *f.* ghrianleictreach
solar-thermal propulsion tiomáint *f.* ghrianteirmeach
solid-earth tide taoide *f.* dhlúthdhomhain
solid propellant dlúth-thiomántán *m.*
solutal convection comhiompar *m.* tuaslagáiteach
solutal Marangoni convection comhiompar *m.* Marangoni tuaslagáiteach
solution transport iompar *m.* tuaslagáin
somatic sensation mothúchán *m.* sómach
sorption sú *m.*
sounding sondáil *f.*
sounding balloon balún *m.* sondála
sounding LiDAR líodar *m.* sondála
sounding radar radar *m.* sondála
sounding rocket roicéad *m.* sondála
sounding rocket experiment turgnamh *m.* roicéid sondála
South Atlantic Anomaly domain [SAA domain] fearann *m.* Aimhrialtacht an Atlantaigh (fearann *m.* SAA)
space spás *m.*
space activities gníomhachtaí *fpl.* spáis
space adaptation syndrome siondróm *m.* spásoiriúntais; siondróm *m.* oiriúntais spáis
space anemia spásainéime *f.*; ainéime *f.* spáis
space architect spásailtire *m.*
space architecture spásailtireacht *f.*
space art spásealaín *f.*; ealaín *f.* spáis
space astronomy réalteolaíocht *f.* spáis
space base bunáit *f.* spáis

space biology bitheolaíocht *f.* spáis
space capsule spáschapsúl *m.*
space colony spáschoilíneacht *f.*
space communications spáschumarsáid *f.*
space debris smionagar *m.* spáis
space device gléas *m.* spáis; feiste *f.* spáis
space discovery fionnachtain *f.* spáis
space division multiple access ilrochtain *f.* roinnte spáis
space elevator spásardaitheoir *m.*
space environment timpeallacht *f.* spáis
space experiment spásturgnamh *m.*
space exploration taiscéalaíocht *f.* spáis
space factory spásmhonarcha *f.*
space flight spáseitilt *f.*
space industry tionscal *m.* spáis
space law dlí *m.* an spáis
space medicine míochaine *f.* spástaistil
space meteorology spásmheitéareolaíocht *f.*
space mission misean *m.* spáis
space motion sickness tinneas *m.* spághluaisne; tinneas *m.* gluaisne spáis
space navigation spásloingseoireacht *f.*
space object réad *m.* spáis
space observation breathnóireacht *f.* spáis; breathnú *m.* spáis
space operation service seirbhís *f.* oibríochta spáis
space physics fisic *f.* spáis
space platform spásardán *m.*
space policy polasaí *m.* spáis
space power[1] **(energy)** cumhacht *f.* spáis
space power[2] **(country)** Cumhacht *f.* spáis
space probe taiscéalaí *m.* spáis
space program clár *m.* spáis
space project tionscadal *m.* spáis
space psychology and psychiatry síceolaíocht *f.* agus síciatracht spáis
space radiocommunications radachumarsáid *f.* spáis
space remote sensing cianbhraiteacht *f.* spáis
space research taighde *m.* spáis
space resources acmhainní *fpl.* spáis
space robot spásróbat *m.*

space segment deighleog *f.* spáis
space ship spáslong *f.*
space shuttle tointeálaí *m.* spáis; spástointeálaí *m.*
space sickness tinneas *m.* spáis
space simulation chamber seomra *m.* ionsamhlaithe spáis; cuasán *m.* ionsamhlaithe spáis
space simulator ionsamhlóir *m.* spáis
space station stáisiún *m.* spáis
space structure struchtúr *m.* spáis
space system córas *m.* spáis
space tourism spásturasóireacht *f.*; turasóireacht *f.* spáis
space tourist spásturasóir *m.*
space traffic spástrácht *m.*
space traffic management bainistíocht *f.* spástráchta
space transportation spásiompar *m.*
space transportation system [STS] córas *m.* iompair spáis
space tug spástuga *m.*
space utilization spásúsáid *f.*
space vacuum folús *m.* spáis
space vehicle feithicil *f.* spáis
space weather spásaimsir *f.*
space weather report faisnéis *f.* spásaimsire
space weathering síonchaitheamh *m.* spáis
space-air vehicle feithicil *f.* spásaeir
spaceborne spásbheirthe
spacecraft spásárthach *m.*
spaceflight participant rannpháirtí *m.* spáseitilte
space-rate (to) spásrátáil
space-rating spásrátáil *f.*
spacesuit culaith *f.* spáis
spacewalks siúlóidí *fpl.* spáis
spaceworthy spásacmhainneach
spatial spásúil
spatial orientation treoshuíomh *m.* spásúil
spatial resolution taifeach *m.* spásúil
spatial sense mothú *m.* spásúil
specific consumption sainídiú *m.*
specific energy sainfhuinneamh *m.*

specific impulse sainríog *f.*
specimen eiseamal *m.*
speckle breacadh *m.*; breac *m.*
spectral emittance astaíocht *f.* speictreach
spectral radiance radaíocht *f.* speictreach
spectral reflectance frithchaiteas *m.* speictreach
spectral resolution taifeach *m.* speictreach
spectral signature síniú *m.* speictreach
spectrometer speictriméadar *m.*
spectroscopic binary star déréalta *f.* speictreascópach
spectrum speictream *m.*
specular reflection frithchaitheamh *m.* spéaclach
sphere of activity réimse *m.* gníomhaíochta
sphere of influence réimse *m.* tionchair
spherical albedo ailbéadacht *f.* sféarúil
spicule spícín *m.*
spill-over[1] (**overflow**) doirteadh *m.* thairis
spill-over[2] (**the excess part of something**) farasbarr *m.*
spin casadh *m.*; guairne *f.*
spin (**to**) casadh
spin stabilization cobhsú *m.* casaidh
spin table tábla *m.* casaidh
spin thruster ceartaitheoir *m.* casaidh
spin-down casadh *m.* síos
spinner casadóir *m.*
spin-off toradh *m.*; seachthairbhe *f.*
spin-up casadh *m.* suas
spin-up (**to**) casadh suas
spiral galaxy réaltra *m.* bíseach
splashdown stealltuirlingt *f.*
split fairing slíomú *m.* scoilte
sponsor urra *m.*; urraitheoir *m.*
spore survival marthain *f.* spóir
spot beam antenna aeróg *f.* spotléis
spot noise temperature teocht *f.* spot-torainn
spread-spectrum speictream *m.* leata
spurious signal comhartha *m.* bréagach
sputtering spiúchadh *m.*

squib pléascáinín *m.*
stabilization cobhsú *m.*
stabilized platform ardán *m.* cobhsaithe
stabilizer cobhsaitheoir *m.*
stage céim *f.*
stage integrator suimeálaí *m.* céime
staging céimniú *m.*
stagnation point pointe *m.* marbhántachta
stagnation pressure brú *m.* marbhántachta
stagnation temperature teocht *f.* marbhántachta
stand-alone neamhspleách
standard atmosphere atmaisféar *m.* caighdeánach
standard time am *m.* caighdeánach
standby redundancy iomarcaíocht *f.* cúltaca
standing wave ratio [**SWR**] cóimheas *m.* toinne seasta
staple stápla *m.*
star réalta *f.*
star formation rate [**SFR**] ráta *m.* réaltfhoirmithe
star tracker rianaire *m.* réalta
starburst galaxy réaltra *m.* rois réalta
starch-statolith hypothesis hipitéis *f.* statailite stáirse
star-forming region réigiún *m.* réaltfhoirmithe
start-stop dúiseacht-stop *m.*
static receptor gabhdóir *m.* statach
static split scoilt *f.* statach
static test tástáil *f.* statach
static test facility saoráid *f.* tástála stataí
static testing tástáil *f.* statach
station stáisiún *m.*
stationary satellite satailít *f.* chónaitheach
station-keeping coinneáil *f.* suímh; coinneáil *f.* stáisiúin
station-keeping satellite satailít *f.* coinneála suímh; satailít *f.* coinneála stáisiúin
statocyst stataicist *f.*
statocyte stataicít *f.*
statolith statailit *f.*
statoreceptor stataghabhdóir *m.*
steerable radiators radaitheoirí *mpl.* instiúrtha

steering stiúradh *m.*
stellar atmosphere atmaisféar *m.* réaltach
stellar collapse imphléascadh *m.* réaltach
stellar magnitude méid *f.* réaltach
stellar seismology seismeolaíocht *f.* réaltach
stellar wind gaoth *f.* réaltach
stepping down of an orbit íoschéimniú *m.* fithise
sterilization of space objects steiriliú *m.* réad spáis; steiriliúchán *m.* réad spáis
stiffener dochtóir *m.*
stony iron meteorite dreigít *f.* iarainn chlochach
stony meteorite dreigít *f.* chlochach
storable propellant tiomántán *m.* instórála
storage stóras *m.*
storage cell cill *f.* stórais
storm shelter scáthlán *m.* stoirme
stowage stuálas *m.*
stowage module modúl *m.* stuálais
strap strapa *m.*
strap-on booster treiseoir *m.* strapáilte
stratopause stratastad *m.*
stratosphere strataisféar *m.*
striation stríocadh *m.*
stringer maide *m.* ceangail
strip stiall *f.*
stripline stiall-líne *f.*
Stroemgren radius raidias *m.* Stroemgren; ga *m.* Stroemgren
Stroemgren sphere sféar *m.* Stroemgren
structural ratio cóimheas *m.* struchtúrach
structure struchtúr *m.*
subcontractor fochonraitheoir *m.*
subgiant star réalta *f.* fofhathaigh; fofhathach *m.*
subgravity fo-imtharraingt *f.*
subjective vertical ingear *m.* suibiachtúil
sublimation néalú *m.*; néalúchán *m.*
submerged nozzle soc *m.* báite
submillimeter astronomy réalteolaíocht *f.* fhomhilliméadair
submillimeter wave tonn *f.* fhomhilliméadair

suborbital flight eitilt *f.* fho-fhithiseach
suborbital trajectory ruthag *m.* fo-fhithiseach
subsatellite foshatailít *f.*
subsatellite point pointe *m.* foshatailíte
subsonic foshonach
subsonics foshonaic *f.*
subsynchronous satellite satailít *f.* fhoshioncronach
subsystem fochóras *m.*
sudden ionospheric burst [**SIB**] corraíl *f.* thobann ianaisféarach
suite svuít *f.*; foireann *f.*
Sun An Ghrian *f.*
sun sensor braiteoir *m.* gréine
sunspot grianspota *m.*
sunspot number líon *m.* na ngrianspotaí
sunspot relative number uimhir *f.* choibhneasta grianspota
sun-synchronous orbit fithis *f.* ghriansioncronach
sun-synchronous satellite satailít *f.* ghriansioncronach
Sunyaev-Zeldovich effect éifeacht *f.* Sunyaev-Zeldovich
supercluster of galaxies ollbhraisle *f.* réaltraí
supercooling sárfhuarú *m.*
supercritical state staid *f.* shárchriticúil
supergalaxy plane plána *m.* ollréaltra
supergiant star réalta *f.* ollfhathaigh; ollfhathach *m.*
supergranulation ollghránú *m.*
superior conjunction cónasc *m.* uachtair
supermassive black hole dúpholl *m.* ollmhór
supernova ollnóva *m.*
supernova remnants fuílleach *m.* ollnóva
super-pressure forbhrú *m.*
super-pressure balloon balún *m.* forbhrú
supersonic forshonach
supersonic-combustion ramjet scairdinneall *m.* dó fhorshonaigh
supersonics forshonaic *f.*
supersynchronous satellite satailít *f.* fhorshioncronach
supervision maoirseacht *f.*
surface dromchla *m.*
surface acoustic wave [**SAW**] tonn *f.* fhuaimiúil dromchla
surface tension teannas *m.* dromchla

surface vibration creathadh *m.* dromchla
survival kit trealamh *m.* marthana
sustainer (engine) cothaire *m.* (inneall)
swath sraith *f.*
swath width leithead *m.* sraithe
sweat cooling fuarú *m.* allais
swing-by luascadh *m.* thart
switching lascadh *m.*; lascachán *m.*
swiveling nozzle soc *m.* sclóinithe
symbiotic star réalta *f.* shiombóiseach
synchronized satellite satailít *f.* shioncronaithe
synchronous satellite satailít *f.* shioncronach
synchrotron radiation radaíocht *f.* sioncrotróin
synodic period tréimhse *f.* shionadach
synthetic aperture radar [SAR] radar *m.* cró shintéisigh
system córas *m.*
system definition phase pas *m.* sainithe córais
system effectiveness éifeachtacht *f.* córais
system hazard guais *f.* chórais
systems architect ailtire *m.* córas
systems engineer innealtóir *m.* córas
systems engineering innealtóireacht *f.* córas
systems sistéamach

T

T Tauri star [TTS] réalta *f.* T Tauri
tactical air navigation aerloingseoireacht *f.* oirbheartaíochta
tail current sruth *m.* deiridh
tail fin cobhsaitheoir *m.* deiridh
tail suspension crochadh *m.* eireabaill
tail unit aonad *m.* deiridh
take-off éirí *m.* de thalamh
targeted search cuardach *m.* sprice
telecommand [tc] teilea-ordú *m.*
telecommunication network líonra *m.* teileachumarsáide
telecommunications teileachumarsáid *f.*
telecommunications satellite satailít *f.* teileachumarsáide
telematics teileamataic *f.*
telemedicine teilileigheas *m.*
telemetry [TM] teiliméadracht *f.*
teleoperation teilea-oibriú *m.*
telepresence teilealáithreacht *f.*
telescience teilea-eolaíocht *f.*
tender tairiscint *f.*
tender evaluation board bord *m.* measúnaithe tairiscinte
tender opening board bord *m.* oscailte tairiscinte
tensegrity sláine *f.* theannasach
terminator solasteorainn *f.*
terms of reference téarmaí *mpl.* tagartha
terraforming teireafoirmiú *m.*; domhanfhoirmiú *m.*
terramechanics meicnic *f.* thalún
terrestrial longitude fad *m.* domhanda
terrestrial planet pláinéad *m.* domhanda
terrestrial station stáisiún *m.* domhanda
test tástáil *f.*
test article earra *m.* tástála
test facility saoráid *f.* tástála

test specimen eiseamal *m.* tástála
tethered balloon balún *m.* ar teaghrán
tethered satellite satailít *f.* ar teaghrán
theory of general relativity teoiric *f.* na coibhneastachta ginearálta
theory of special relativity teoiric *f.* na coibhneastachta speisialta
thermal accommodation coefficient comhéifeacht *f.* oiriúna theirmeach
thermal balance cothromaíocht *f.* theirmeach
thermal balloon balún *m.* teirmeach
thermal barrier bacainn *f.* theirmeach
thermal control rialú *m.* teirmeach
thermal convection comhiompar *m.* teirmeach; comhiompar *m.* teasa
thermal efficiency éifeachtacht *f.* theirmeach
thermal protection cosaint *f.* theirmeach
thermal screen scáileán *m.* teirmeach
thermal shock turraing *f.* theirmeach
thermoluminescence dosimeters [TLD] dáileogmhéadair *mpl.* theirmealonrachta
thermo-nuclear teirmeanúicléach
thermopile teirmeacharn *m.*
thermo-regulation teirmirialú *m.*
thermo-solutal convection comhiompar *m.* teirmeathuaslagáiteach
thermosphere teirmisféar *m.*
third cosmic velocity an tríú treoluas *m.* cosmach
three-axis stabilization cobhsaíocht *f.* trí ais
three-axis stabilized cobhsaithe ag trí ais
three-body problem fadhb *f.* trí chorp
three-dimensional clinostat clíneastat *m.* tríthoiseach
3kpc arm géag *f.* 3kpc
throat scornach *f.*
thrust sá *m.*
thrust axis ais *f.* sá
thrust balance cothromaíocht *f.* sá
thrust coefficient comhéifeacht *f.* sá
thrust control command rialú *m.* na n-orduithe sá
thrust cut-off scoitheadh *m.* sá
thrust decay meath *m.* sá
thrust frame fráma *m.* sá
thrust misalignment mí-ailíniú *m.* sá

thrust terminator críochtóir *m.* sá
thrust vector control [**TVC**] rialú *m.* veicteora sá
thruster ceartaitheoir *m.*
Thule group grúpa *m.* Thule
tidal action gníomhú *m.* taoidmhear
tidal component comhbhall *m.* taoidmhear
tidal forces fórsaí *mpl.* taoide
tidal friction frithchuimilt *f.* thaoidmhear
tidal torque torc *m.* taoidmhear
time and material contract conradh *m.* ama agus ábhair
time delay and integration device [**TDI**] gléas *m.* suimeála agus moille ama; feiste *f.* suimeála agus moille ama
time division multiple access [**TDMA**] ilrochtain *f.* roinnte ama
time division multiplexing [**TDM**] ilphléacsú *m.* roinnte ama
time of perihelion passage am *m.* shroicheadh peiriheilin
time resolution taifeach *m.* ama
time sharing comhroinnt *f.* ama
Tisserand's criterion critéar *m.* Tisserand
Titan Tíotán *m.*
Titius-Bode's law dlí *m.* Titius-Bode
tone ton *m.*
tone ranging raonáil *f.* toin
top up (**to**) barrlíonadh
topocentric angle uillinn *f.* thopalárnach
topological defect fabht *m.* toipeolaíoch
topping up barrlíonadh *m.*
topside (**ionospheric**) **sounding** sondáil *f.* (ianaisféarach) uachtair
total emissivity lánastaíochas *m.*
total impulse lánríog *f.*
total pressure lánbhrú *m.*
touch screen scáileán *m.* tadhaill
tracker rianaire *m.*
tracking rianú *m.*
tracking and data relay satellite satailít *f.* rianaithe agus athsheachadta sonraí
tracking antenna aeróg *f.* rianaithe
tracking camera ceamara *m.* rianaithe
tracking facilities saoráidí *fpl.* rianaithe

tracking radar radar *m.* rianaithe
tracking station stáisiún *m.* rianaithe
tracking telemetry and command [TT&C] [TTC] rianú *m.* teilimeádracht agus rialú
trade-off comhbhabhtáil *f.*
traffic control rialú *m.* tráchta
trailing edge imeall *m.* deiridh
train sraith *f.*
trajectory ruthag *m.*
trajectory determination cinneadh *m.* ruthaig
transceiver trasghlacadóir *m.*
transcoder traschódóir *m.*
transcoding traschódú *m.*
transducer trasduchtóir *m.*
transfer orbit fithis *f.* aistrithe
transfer stage céim *f.* aistrithe
transhorizon propagation forleathadh *m.* trasléaslíne
transient acceleration luasghéarú *m.* díomuan
transient hot-wire method modh *m.* teoshreangach díomuan
transionospheric propagation forleathadh *m.* trasianaisféarach
transit trasdul *m.*
transit circle ciorcal *m.* trasdula
transit instrument ionstraim *f.* thrasdula
transmission channel cainéal *m.* tarchuir
transmission coefficient comhéifeacht *f.* tarchuir
transmittance tarchuras *m.*
trans-Neptunian object [TNO] réad *m.* tras-Neiptiúnach
transonic tarshonach
transonics tarshonaic *f.*
transpiration cooling fuarú *m.* trasghalúcháin
transponder trasfhreagróir *m.*
transponder beacon rabhchán *m.* trasfhreagróra
transportation in space iompar *m.* sa spás
trapped particle cáithnín *m.* gafa
traveling wave tube [TWT] feadán *m.* toinne taistil
treadmill training traenáil *f.* mhuileann coise
tri-propellant tré-thiomántán *m.*
Trojan asteroid astaróideach *m.* Traíoch

Trojan planet pláinéad *m.* Traíoch
tropical cyclone cioclón *m.* trópaiceach; cioclón *m.* teochriosach
tropical year bliain *f.* na gréine; bliain *f.* thrópaiceach
tropism trópacht *f.*
tropopause trópastad *m.*
troposphere trópaisféar *m.*
tropospheric propagation forleathadh *m.* trópaisféarach
true albedo ailbéideacht *f.* fhíor
true anomaly aimhrialtacht *f.* fhíor
true eclipse urú *m.* fíor
true solar time grianam *m.* fíor
truncated test tástáil *f.* theasctha; tástáil *f.* bharrscoite
truss trus *m.*
truss assembly cóimeáil *f.* trus
truss structure struchtúr *m.* trus
Tully-Fisher relation coibhneas *m.* Tully-Fisher
tumbling timbleáil *f.*
turbine wheel roth *m.* tuirbín
turbo-fan turbafean *m.*
turbo-jet inneall *m.* turbascairde
turbo-pause turbastad *m.*
turbo-pump turbacaidéal *m.*
turbulence suaiteacht *f.*
turbulent boundary layer sraith *f.* teorann suaite
turbulent flow sreabhadh *m.* suaite; sreabhadh *m.* sruthlamach
turnaround time (equipment) aga *m.* slánúcháin (trealamh)
turnover láimhdeachas *m.*; athrúchán *m.*
twilight breacsholas *m.*; clapsholas *m.*
twin paradox paradacsa *m.* cúpla
two liquid phase separation pasdeighilt *f.* dhá leacht
two-phase flow sreabhadh *m.* déphasach; sreabh *f.* dhéphasach
typhoon tíofún *m.*

U

ullage ulláiste *m.*

ultra-high vacuum folús *m.* ultra-ard

ultra-luminous infrared galaxy réaltra *m.* infridhearg ultralonrúil

ultra-luminous X-ray source foinse *f.* X-ghathach ultralonrúil

ultrasonic level sensor braiteoir *m.* leibhéil ultrasonaigh

ultra-stable oscillator [USO] ascaltóir *m.* ultrachobhsaí; ascaltóir *m.* sárchobhsaí

ultraviolet astronomy réalteolaíocht *f.* ultravialait

umbilical imleacánach

umbilical cable cábla *m.* imleacáin

umbilical connector cónascaire *m.* imleacáin

umbilical cord sreang *f.* imleacáin

umbilical mast crann *m.* imleacáin

umbra lánscáil *f.*

umbrella reflector antenna aeróg *f.* frithchaiteora scátha

under-expanded fofhairsingthe

under-expansion fofhairsingiú *m.*

underground faoi thalamh

undisturbed orbit fithis *f.* neamhshuaite

undocking dí-nascadh *m.*

unfurlable antenna aeróg *f.* inscaoilte

unidentified flying object [UFO] réad *m.* eitilte gan aithint; úfó *m.*

unidentified infrared band banda *m.* infridhearg gan aithint

universal uilíoch

universal gravitation imtharraingt *f.* uilíoch

universal time coordinated [UTC] am *m.* uilíoch lárnach [AUL]

universal time [UT] am *m.* uilíoch

universality uilíocht *f.*

Universe An Chruinne *f.*

unlocking díghlasáil *f.*

unmanned flight eitilt *f.* gan fhoireann

unmanned mission misean *m.* gan fhoireann

unmanned spacecraft spásárthach *m.* gan fhoireann
unperturbed orbit fithis *f.* gan chorraíl
unpriming díphrímeáil *f.*
unstable growth fás *m.* éagobhsaí
unsteady Marangoni convection comhiompar *m.* Marangoni neamhfhoisteanach
uplink uasnasc *m.*
upper atmosphere atmaisféar *m.* uachtarach
upper stage céim *f.* uachtarach
Uranus Úranas *m.*
use of outer space úsáid *f.* an spáis amuigh
useful life saolré *f.* fhónta
utricle úitrín *m.*

V

Van Allen belt crios *m.* Van Allen
variable conductance heat pipe píobán *m.* teasa le seoltas athraitheach
variable star réalta *f.* athraitheach
vasoconstriction caolú *m.* fuileadán
Vega-like star réalta *f.* mar Vega
vehicle equipment bay bá *f.* trealaimh feithicile
velocimetry luasmhéadracht *f.*
veno-arterial reflex athfhilleadh *m.* féithartaireach
venoconstriction caolú *m.* féithe
vent gaothaire *m.*
vent duct ducht *m.* gaothaire
vent hole poll *m.* gaothaire
vent tube feadán *m.* gaothaire
ventilation distribution dáileadh *m.* aerála
venturi feadán *m.* Venturi; éifeacht *f.* Venturi
Venus Véineas *f.*
verification fíorú *m.*; fíorúchán *m.*
vernal equinox cónocht *m.* an earraigh
vernal point pointe *m.* earraigh
Vernier engine inneall *m.* Vernier
Vernier motor mótar *m.* Vernier
versatile ilúsáideach
versatility ilúsáideacht *f.*
vertical (ionospheric) sounding sondáil *f.* (ianaisféarach) cheartingearach
vertical reflectivity frithchaiteacht *f.* cheartingearach
vertical stabilizer cobhsaitheoir *m.* ceartingearach
vertigo meadhrán *m.*
very long baseline interferometry [VLBI] trasnamhéadracht *f.* bunlíne an-fhada
vestibular compensation cúiteamh *m.* cuasánach
vestibular nucleus núicléas *m.* cuasánach

vestibular organ ball *m.* cuasánach; orgán *m.* cuasánach
vestibular otolith otailit *f.* chuasánach
vestibular system córas *m.* cuasánach
vestibular test tástáil *f.* chuasánach
vestibulo-ocular reflexes athfhillteacha *mpl.* cuasánacha súilí
vestibulo-spinal reflexes athfhillteacha *mpl.* cuasánacha dromlaigh
vestibulo-spinal response freagairt *f.* chuasánach dromlaigh
vibration damper maolaire *m.* creathaidh
vibration environment timpeallacht *f.* chreathaidh
vibration generator gineadóir *m.* creathaidh
vibration isolation device [VID] gléas *m.* aonraithe creathaigh; feiste *f.* aonraithe creathaigh
vibrator creathóir *m.*; creathadóir *m.*
videofrequency [VF] físmhinicíocht *f.*
viewing angle uillinn *f.* féachana
virial equilibrium cothromaíocht *f.* viriúil
visible arc stua *m.* infheicthe
visual acuity géire *f.* amhairc
visual binary dénarthán *m.* amhairc
visual display unit [VDU] aonad *m.* taispeána físe
visual magnitude amharcmhéid *f.*
visual-vestibular interaction idirghníomhaíocht *f.* amharc-chuasánach
volatile element eilimint *f.* sho-ghalaithe
voltage controlled oscillator [VCO] ascaltóir *m.* rialaithe ag voltas
voltage standing wave ratio [VSWR] cóimheas *m.* toinne seasta voltais
volumetric efficiency éifeachtacht *f.* thoirtmhéadrach
vortex cuilithe *f.*

W

wake marbhshruth *m.*
wall temperature teocht *f.* balla
warhead pléasc-cheann *m.*
waste management bainistíocht *f.* dramhaíola
water hole poll *m.* uisce
water immersion tumadh *m.* uisce
water maser méasar *m.* uisce
water-activated battery ceallra *m.* uisce-ghníomhachtaithe
wave interference trasnaíocht *f.* toinne
weather radar radar *m.* aimsire
weight-bearing muscle matán *m.* seasamh meáchain
weighting factor fachtóir *m.* ualaithe
weightlessness dímheáchan *m.*
whip antenna aeróg *f.* fuipe
whistler feadaire *m.*
whistler mode mód *m.* feadaire
white dwarf star abhacréalta *f.* bhán
white noise torann *m.* bán
wick (heat pipe) buaiceas *m.*; fáideog *f.* (píobán teasa)
Widmanstatten structure struchtúr *m.* Widmanstatten
wind tunnel tollán *m.* gaoithe
wind-tunnel balance cothromaíocht *f.* tolláin gaoithe
winding tochrán *m.*
wing sciathán *m.*
wing load ualach *m.* sciatháin
wing profile próifíl *f.* sciatháin
wing section gearradh *m.* sciatháin
work breakdown structure [WBS] struchtúr *m.* miondealaithe oibre
work package pacáiste *m.* oibre
working fluid sreabhán *m.* oibre
world weather watch [WWW] faire *f.* aimsire an domhain
worldwide domhanda; ar fud an domhain
worldwide communications cumarsáid *f.* dhomhanda

X

X-ray astronomy réalteolaíocht *f.* X-gha
X-ray bright point gealphointe *m.* X-gha
X-ray burst rois *f.* X-gha
X-ray halo luan *m.* X-gha
X-ray star réalta *f.* X-gha
X-ray telescope teileascóp *m.* X-gha
x-y mount taca *m.* x-y

Y

Yagi antenna aeróg *f.* Yagi
Yarkovsky effect éifeacht *f.* Yarkovsky
yaw luascáil *f.*
year bliain *f.*
yoyo yóyó *m.*

Z

zenith forar *m.*

zenith distance fad *m.* forair

zero gravity imtharraingt *f.* niallais; domhantarraingt *f.* niallais

zero-g g-niallasach

zodiac stoidiaca *m.*

zodiacal counterglow frithloinnir *f.* stoidiacach

zodiacal light solas *m.* stoidiacach

zone són *m.*

zoom[1] (**move rapidly**) súmáil *f.*

zoom[2] (**climb suddenly and sharply**) géarardú *m.*